From Utopian to Unconventional Computers

Splendeurs et misères du calcul peu usuel

Andrew Adamatzky and Christof Teuscher
Editors

From Utopian to Genuine Unconventional Computers

Splendeurs et misères du calcul peu usuel

Luniver Press
2006

Published by Luniver Press
Frome BA11 6TT United Kingdom

British Library Cataloguing-in-Publication Data
A catalogue record for this book is available from the British Library

From Utopian to Genuine Unconventional Computers: Splendeurs et misères du calcul peu usuel

ISBN-10: 0-9551170-9-7
ISBN-13: 978-0-9551170-9-1

1st edition

Editorial

"... it is stimulating to realize the mathematical conditions of the problem by corresponding physical devices, or rather, to consider the mathematical problem as an interpretation of a physical phenomenon. The existence of the physical phenomenon then represents the solution of the mathematical problem" — this is how, back in 1941, Richard Courant and Herbert Robbins[1] outlined, yet not explicitly named unconventional computing. Nowadays *unconventional computing* includes, but is not limited to physics of computation, chemical and bio-molecular computing, cellular automata, non-classical logics, amorphous computing, and mechanical computation.

It is a vibrant and interdisciplinary field aimed at developing new algorithmic paradigms and novel physical computing substrates. While the last twenty years were mainly dominated by abstract unconventional machines, there is an increasing effort in building genuine realizations. Compared to simulations of abstract machines, genuine machines generally allow to fully and directly exploit the computational capabilities — including the usually massive parallelism — of the physics or chemistry of the underlying substrate.

In organising the 2006 workshop *From Utopian to Genuine Unconventional Computers: Splendeurs et misères du calcul peu usuel*[2], we were eager to resurrect decades old ideas on how to embed and to do computation in physical substrates. We wanted to bring to light working prototypes of unconventional computers akin to Riemann's experiments on the flow of electricity in thin metallic sheet, akin to Lord Kelvin's analog differential analyzer, and akin to Plateau's experiments on the calculation of the surface of the smallest area bounded by a given closed contour. We did not exactly get what we were hoping for, nevertheless, the nine contributed papers to this volume are exciting enough to be flagships of the field. Our thirst for experimental prototypes and genuine

[1] Courant R. and Robbins H. What is Mathematics? (Oxford University Press, 1941).

[2] The workshop was organized as part of the 5th *International Conference on Unconventional Computation*, UC 2006, which was held Sep 4–8, 2006, at the University of York, UK. We would like to express our gratitude to the members of the workshop program committee, who did a great job in vigorously reviewing every submission and ensuring an exceptional quality of the accepted papers. We're also very grateful to the UC'06 organizers for giving us the opportunity to hold this workshop!

implementations of non-classical computing devices is indeed satisfied by the majority of the contributions in this volume. In addition, we have also decided to include a number of more theoretical papers of particular interest.

In the very first chapter, “On digital VLSI circuits exploiting collision-based fusion gates”, Yamada, Motoike, Asai, and Amemiya propose a novel method to design VLSI circuits, based on fusion gates, which itself was discovered by designing logical circuits in excitable chemical media. They show an efficient way to build logic gates, which would at the same time allow them to operate at high speed and low power.

The ultimate goal of Larry Bull and his collaborators is to build a chemical neural network in Belousov-Zhabotinsky media. The first stage of the project — which consists in controlling and adapting sub-excitable chemical medium — is addressed in their contribution “Towards machine learning control of chemical computers”. They do not employ geometrically-constraint media but use the light-sensitivity of the system to impose dynamical structures of quasi-neuronal connections instead. They then apply machine learning to direct wave fragments to an arbitrary position, and thus to solve a specific task.

Nicolas Glade gives us a biochemist's view on computation. He explores micro-tubules as a media for *in vitro* computation. The computations are based on "chemical collisions" between micro-tubules. No definite computation schemes are provided, however, experimental evidence discussed suggests a huge computational potential with micro-tubules.

Yet another unconventional approach for a potential computing system is presented by Greenman, Ieropolous, and Melhuish: the implementation of Pavlovian reflexes in bacterial films. We envision that techniques outlined in “Perfusion anodophile biofilm electrodes and their potential for computing” will be used in future designs of bacteria-based controllers for autonomous robots, which are powered by bacterial cells.

Finally, in the last more practically-oriented contribution, Murphy et al. implement physical sorting by using various methods, such as gel electrophoresis, mass spectroscopy, and chromatography. Amongst other wonderful things, their paper “Implementations of a model of physical sorting” tells us again that computation in physics and Nature is ubiquitous.

Universal computation is a key concept in computer science since its beginnings in the early fifties. In his contribution “Conventional or unconventional: Is any computer universal?”, Selim Akl discusses different evolving computational paradigms.

Wiesner and Crutchfield explore the language diversity of quantum finite-state generators — which occupy the lowest level of the still partially unknown hierarchy of quantum computation — and show that deterministic quantum finite-state generators have a larger language diversity than their classical analog stochastic finite-state generators. Concepts studied in Wiesner and Crutchfield's "Language diversity of measured quantum processes" are currently rather far from any real-world implementation, however, they do certainly stimulate our brains with the esthetic of theoretical aspects.

Collision-based computing is a well-known unconventional computing paradigm, traced back to Conway's *Game of Life*, Fredkin-Toffoli's conservative logic, and Steiglitz's particle machines. In "Logic circuits in a system of repelling particle", William Stevens develops a kinematic model of movable tiles in a two-dimensional environment and shows how dual-rail logic gates can be constructed.

Finally, De Vos and Van Rentergem study in "From group theory to reversible computer" the building-blocks for reversible computing architectures and show that a particular set can be used to synthesize arbitrary reversible circuits. They have also built and tested silicon implementations of adder circuits.

We certainly hope that our readers will enjoy the wide variety of non-classical computing devices — be it chemical, molecular, electronic, or quantum — presented in this timely volume. The contributions attest that there are almost no limits with regards to what physical substrates are used and how computations are performed.

"I don't think God's a computation. But exactly why not?", Rudy Rucker wrote[3].

June 2006

Andy Adamatzky and Christof Teuscher
Workshop Chairs,
Bristol and Los Alamos

[3] Rudy Rucker, The Lifebox, The Seashell, and The Soul. Thunder's Mouth Press, New York, 2006.

Referees

Andrew Adamatzky, University of the West of England, UK
Tetsuya Asai, Hokkaido University, Japan
Ben de Lacy Costello, University of the West of England, UK
Peter Dittrich, Friedrich-Schiller-University Jena, Germany
Jerzy Gorecki, Polish Academy of Science, Poland
Norman Margolus, MIT Artificial Intelligence Laboratory, USA
Genaro Martinez, National University of Mexico, Mexico
Julian F. Miller, University of York, UK
Jonathan W. Mills, Indiana University, USA
Kenichi Morita, Hiroshima University, Japan
Ikuko Motoike, Future University Hakodate, Japan
Ferdinand Peper, Kansai Advanced Research Center, Japan
Ken Steiglitz, Princeton University, USA
Susan Stepney, University of York, UK
Christof Teuscher, Los Alamos National Laboratory, USA
Tommaso Toffoli, Boston University, USA
Hiroshi Umeo, Osaka Electro-Communication University, Japan
Damien Woods, University College Cork, Ireland
Thomas Worsch, Universitaet Karlsruhe, Germany
Kenichi Yoshikawa, Kyoto University, Japan
Klaus-Peter Zauner, University of Southampton, UK

Table of Contents

From Utopian to Genuine Unconventional Computers

On digital VLSI circuits exploiting collision-based fusion gates

Kazuhito Yamada[1], Tetsuya Asai[1], Ikuko N. Motoike[2], and Yoshihito Amemiya[1]

[1] Graduate School of Information Science and Technology, Hokkaido University,
Kita 14, Nishi 9, Kita-ku, Sapporo, 060-0814 Japan
[2] Future University - Hakodate, 116-2 Kamedanakano-cho Hakodate Hokkaido, 041-8655 Japan

Abstract. Collision-based reaction-diffusion computing (RDC) represents information quanta as traveling chemical wave fragments on an excitable medium. Although the medium's computational ability is certainly increased by utilizing its spatial degrees of freedom [2], our interpretation of collision-based RDC in this paper is that wave fragments travel along 'limited directions' 'instantaneously' as a result of the 'fusion of particles'. We do not deal with collision-based computing here, but will deal with conventional silicon architectures of a 'fusion gate' inspired by collision-based RDC. The hardware is constructed of a population of collision points, i.e., fusion gates, of electrically equivalent wave fragments and physical wires that connect the fusion gates to each other. We show that i) fundamental logic gates can be constructed by a small number of fusion gates, ii) multiple-input logic gates are constructed in a systematic manner, and iii) the number of transistors in specific logic gates constructed by the proposed method is significantly smaller than that of conventional logic gates while maintaining high-speed and low-power operations.

1 Introduction

Present digital VLSI systems consist of a number of combinational and sequential logic circuits as well as related peripheral circuits. A well-known basic logic circuit is a two-input NAND circuit that consists of four metal-oxide semiconductor field-effect transistors (MOS FETs) where three transistors are on the current path between the power supply and the ground. Many complex logic circuits can be constructed by not only populations of a large number of NAND circuits but also special logic circuits with a small number of transistors (there are more than three transistors on the current path) compared with NAND-based circuits.

A straight-forward way to construct low-power digital VLSIs is to decrease the power-supply voltage because the power consumption of digital circuits is proportional to the square of the supply voltage. In complex logic circuits, where many transistors are on the current paths, the supply voltage cannot be decreased due to stacking effects of transistors' threshold voltages, even though the threshold voltage is decreasing as LSI fabrication technology advances year by year. On the other hand, if two-input basic gates that have the minimum number of transistors (three or less) on the current path are used to decrease the supply voltage, a large number of the gates will be required for constructing complex logic circuits.

The Reed-Muller expansion [13, 14], which expands logical functions into combinations of AND and XOR logic, enables us to design 'specific' arithmetic functions with a small number of gates, but it is not suitable for arbitrary arithmetic computation. Pass-transistor logic (PTL) circuits use a small number of transistors for basic logic functions but additional level-restoring circuits are required for every unit [18]. Moreover, the acceptance of PTL circuits into mainstream digital design critically depends on the availability of tools for logic, physical synthesis, and optimization. Current-mode logic circuits also use a small number of transistors for basic logic, but their power consumption is very high due to the continuous current flow in turn-on states [5]. Subthreshold logic circuits where all the transistors operate under their threshold voltage are expected to exhibit ultra-low power consumption, but the operation speed is extremely slow [16, 17]. Binary decision diagram logic circuits are suitable for next-generation semiconductor devices such as single-electron transistors [6, 15, 20], but not for present digital VLSIs because of the use of PTL circuits.

To address the problems above concerning low-power and high-speed operation in digital VLSIs, we describe a method of designing logic circuits with collision-based fusion gates, which is inspired by collision-based reaction-diffusion computing (RDC) [2,11,12]. This paper is organized as follows. In section 2, we briefly overview collision-based RDC. Then, in section 3, we introduce a new interpretation of collision-based RDC, especially concerning directions and speeds of propagating information quanta. We also show basic logical functions constructed by simple unit operators, i.e., fusion gates, and demonstrate a circuit's operation by using a simulation program with integrated circuit emphasis

(SPICE) with typical device parameters. Then, a reconfigurable architecture for the proposed fusion-gate structure and a possible construction of D-type flip flop circuits for constructing sequential circuits are presented. Section 4 is a summary.

2 Collision-based logical computation

Dynamic, or collision-based, computers employ mobile self-localizations, which travel in space and execute computation when they collide with each other. Truth values of logical variables are represented by the absence or presence of the traveling information quanta. There are no predetermined wires: patterns can travel anywhere in the medium, a trajectory of a pattern motion is analogous to a momentarily wire [1–3]. A typical interaction gate has two input 'wires' (trajectories of colliding mobile localizations) and, typically, three output 'wires' (two 'wires' represent localization trajectories when they continue their motion undisturbed, the third output gives a trajectory of a new localization, formed in the collision of two incoming localizations). The traveling of patterns is analogous to information transfer while collision is an act of computation; thus, we call the set up 'collision-based computing'. There are three sources of collision-based computing: proof of the universality of Conway's Game of Life via collisions of glider streams [7], conservative logic [8], cellular automaton implementation of the billiard ball model [9], and particle machine [19] (a concept of computation in cellular automata with soliton-like patterns); see overviews in [2].

The main purpose of collision-based computing is to perform computation in an 'empty space', i.e., a medium without geometrical constraints. Basic toy models of collision-based computing are shown in Fig. 1. In the billiard ball logic shown in Fig. 1(a), a set of billiard balls are fired into a set of immovable reflectors at a fixed speed. As the billiard balls bounce off each other and off the reflectors, they perform a reversible computation. Provided that the collisions between the billiard balls and between the billiard balls and the reflectors are perfectly elastic, the computation can proceed at a fixed finite speed with no energy loss.

Adamatzky demonstrated that a similar computation can be performed on excitable reaction diffusion systems [2, 3]. Figure 1(b) illustrates basic logic gates where instead of billiard ball wave fragments (white localizations in the figure) travel in an excitable reaction-diffusion medium. In typical excitable media, localized wave fragments facing each other disappear when they collide. With a special setup described in [3], those excitable waves do not disappear, but they do produce subsequent excitable waves.

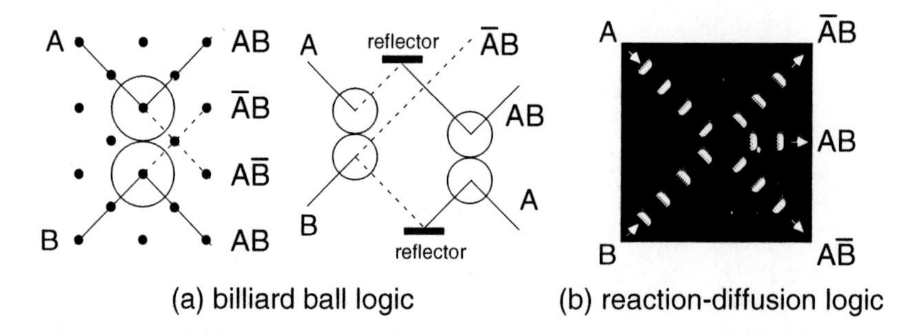

Fig. 1. Collision-based computing models (a) conservative billiard-ball logic for AND and partial XOR computation [8, 9], and (b) nonconservative (dissipative) reaction-diffusion logic that has the same function as that of (a) [3].

3 New interpretation of collision-based computing for digital VLSIs

Adamatzky proposed how to realize arithmetical scheme using wave fragments traveling in a homogeneous medium [2, 3]. The sub-excitable Belousov-Zhabotinsky (BZ) system was emulated by a 2^+-medium [1], which was a 2D cellular automaton, where each cell took three states — resting, excited, and refractory, and updated its state depending on the number of excited cells in its eight-cell neighborhood. A resting cell become excited only if it had exactly two excited neighbors. An excited cell took the refractory state and refractory cell took the resting state unconditionally, i.e., independently of its neighborhood. The model exhibited localized excitations traveling along columns and rows of the lattice and along diagonals. The particles represented values of logical variables. Logical operations were implemented when particles collided and were annihilated or reflected as a result of the collision. Thus one can achieve basic logical gates in the cellular-automaton model of a sub-excitable BZ medium and build an arithmetic circuit using the gates.

A cellular automaton LSI that implements an excitable lattice for BZ systems has been implemented by one of the authors [4, 10]. Each cell consisted of several tens of transistors and was regularly arranged on a 2D chip surface. To implement a one-bit adder, for example, by collision-based cellular automata, at least several tens of cells are required to allocate sufficient space for the collision of wave fragments [2]. This implies several hundreds of transistors are required for constructing just a one-bit adder. Direct implementation of the cellular automaton model is therefore a waste of chip space, as long as the single cell space is decreased to the same degree of chemical compounds in spatially-continuous reaction-diffusion processors.

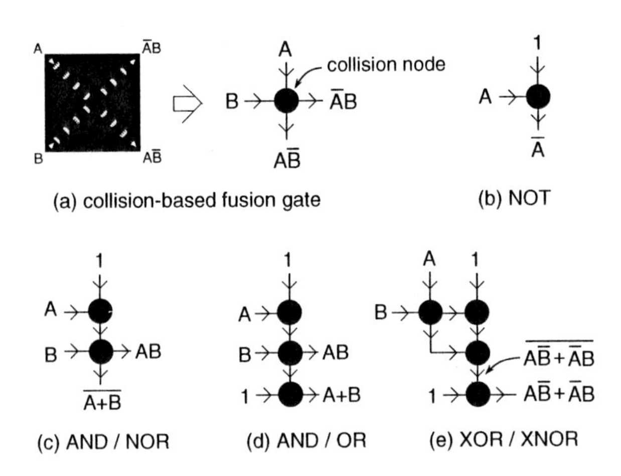

Fig. 2. Definition of collision-based fusion gate (a) and basic logical circuits using several fusion gates [(b)-(e)] that produce multiple logical functions.

What happens if wave fragments travel in limited directions instantaneously? When such wave fragments are generated at the top and end of a pipe (not an empty space) filled with excitable chemicals, for example, these waves may disappear at the center of the pipe instantaneously. When two pipes are perpendicularly arranged and connected, wave fragments generated at the tops of the two pipes may also disappear at the connected point. If only one wave fragment (A or B) is generated at the top of one pipe, it can reach the end of the pipe [$A\bar{B}$ or $\bar{A}B$ in Fig. 1(b)]. A schematic model of this operation is shown in Fig. 2. In Fig. 2(a) (left), an excitable reaction-diffusion medium, where excitable waves (A and B) may disappear when they collide, is illustrated. In Fig. 2(a) (right), an equivalent model of two perpendicular directions of wave fragments, i.e., North-South and West-East fragments, is depicted. The input fragments are represented by values A and B where A (or B) = '1' represents the existence of a wave fragment traveling North-South (or West-East), and A (or B) = '0' represents the absence of wave fragments. When A =

B = '1' wave fragments collide at the center position (black circle) and then disappear. Thus, East and South outputs are '0' because of the disappearance. If A = B = '0', the outputs will be '0' as well because of the absence of the fragments. When A = '1' and B = '0', a wave fragment can travel to the South because it does not collide with a fragment traveling West-East. The East and South outputs are thus '0' and '1', respectively, whereas they are '1' and '0', respectively, when A = '0' and B = '1'. Consequently, logical functions of this simple 'operator' are represented by $\overline{A}B$ and $A\overline{B}$, as shown in Fig. 2(a) (right). We call this operator a 'collision-based fusion gate', where two inputs correspond to perpendicular wave fragments, and two outputs represent the results of collisions (transparent or disappear) along the perpendicular axes. Notice that in this configuration the computation is performed with geometrical constraints. Figures 2(b) to (e) represent basic logic circuits constructed by combining several fusion gates. The simplest example is shown in Fig. 2(b) where the NOT function is implemented by a single fusion gate. The North input is always '1', whereas the West is the input (A) of the NOT function. The output appears on South node ($\overline{A}$). Figure 2(c) represents a combinational circuit of two fusion gates that produces AND and NOR functions. An OR function is thus obtained by combining NOT and AND/NOR fusion gates in Figs. 2(b) and (c), respectively, as shown in Fig. 2(d). Exclusive logic functions are produced by three (for XNOR) or four (for XOR) fusion gates as shown in Fig. 2(e).

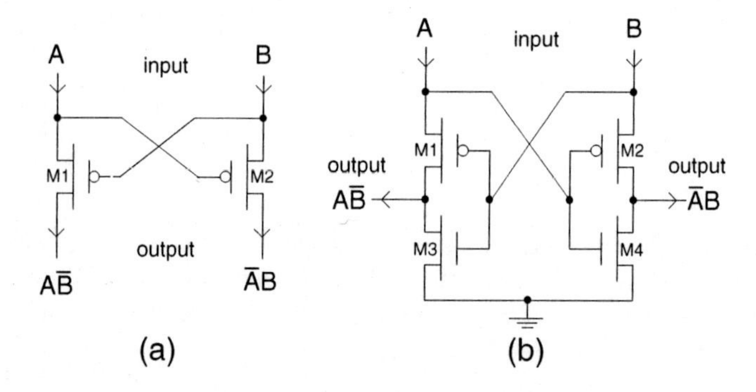

Fig. 3. Circuit construction of collision-based fusion-gate for digital VLSIs. (a) compact-but-slow circuit (two-transistor circuit) and (b) complex-but-fast circuit (four-transistor circuit).

A collision-based fusion gate receives two logical inputs (A and B) and produces two logical outputs ($\overline{A}B$ and $A\overline{B}$). CMOS circuits for this gate are shown in Fig. 3. They receive logical (voltage) inputs (A and B) and produce the logic function. The minimum circuit structure based on PTL circuits is shown in Fig. 3(a), where a single-transistor AND logic is fully utilized. When an nMOS transistor receives voltages A and B at its gate and drain, respectively, the source voltage is given by AB at equilibrium; in the case of pMOS, that is given by $\overline{A}B$. Although there are just two transistors in this construction, there is a severe problem, i.e., the operation speed. When a pMOS transistor is turned off, the output node's parasitic capacitance is discharged by the leak current of the pMOS transistor in the off state. Therefore, the transition time will be rather long, e.g., typically within a few tens of milliseconds when the conventional CMOS process is used, which implies the circuit operates very slowly compared with conventional digital circuits. Additional resistive devices for the discharging may improve the upper bound of the clock frequency. However, we need a breakthrough while satisfying the constraints of a small number of transistors in a fusion gate. One solution to the operation-speed problem is shown in Fig. 3(b). The circuit has two additional nMOS transistors just beneath the pMOS transistors in the two-transistor circuits. If a pMOS transistor is turned off, an nMOS transistor connected between the pMOS transistor and the ground discharges the output node, which significantly increases the upper bound of the operation frequency.

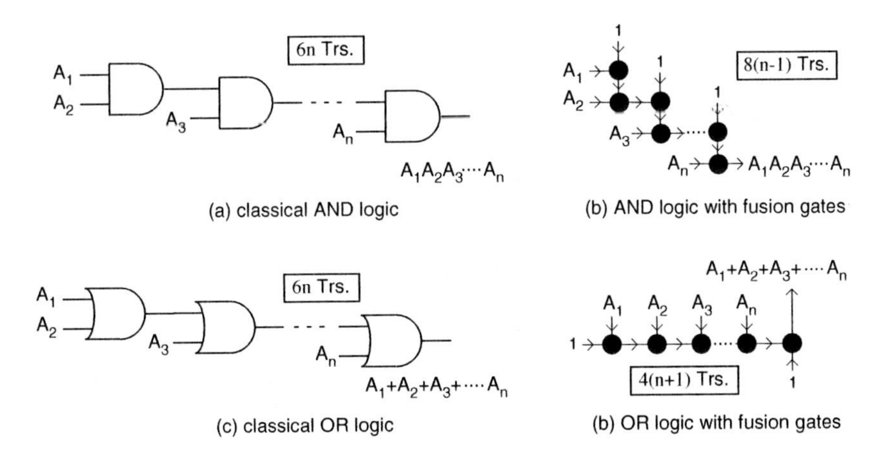

Fig. 4. Classical and fusion-gate logic architectures of multiple-input AND and OR circuits. In classical circuits [(a) and (c)], each logical unit consists of six transistors with constraints for low-power operation; three transistors are on current path between power supply and ground. In fusion-gate logic [(b) and (d)], each fusion gate consists of four transistors as shown in Fig. 3(b).

A multiple-input AND and OR implementation with classical and fusion-gate logic is shown in Fig. 4. In classical circuits (a) and (c), each logical unit (two-input AND and OR) consists of six transistors. As introduced in first section, to decrease the power supply voltage for low-power operation, a minimum number of transistors (three or less) should be on each unit's current path. Each unit circuit has six transistors, so n-input AND and OR gates consist of $6(n-1)$ transistors ($n \geq 2$). On the other hand, in fusion-gate logic (b) and (d), an n-input AND gate consists of $8(n-1)$ transistors [$2(n-1)$ fusion gates], whereas $4(n+1)$ transistors will be used in an n-input OR gate. Therefore, in the case of AND logic, the number of transistors in classical circuits is smaller than that of fusion-gate circuits. However, in the case of OR circuits, the number of transistors in fusion-gate circuits is smaller than that of classical circuits, and the difference will be significantly expanded as n increases.

Classical and fusion-gate implementations of majority logic circuits with multiple inputs are shown in Fig. 5. Again, in classical circuits, the number of transistors on each unit's current path is fixed to three. Five-input majority gates are illustrated in the figure. For n-bit inputs (n must be an odd number larger than 3), the number of transistors in the classical circuit was $30+36(n-3)^2$, while in the fusion-gate circuit, it was $2n(n+1)$, which indicates that the fusion-gate circuit has a significantly smaller number of transistors.

Half and full adders constructed by classical and fusion-gate logic are illustrated in Fig. 6. There were 22 transistors in a classical half adder [Fig. 6(a)], whereas they were was 32 in a fusion-gate half adder [Fig. 6(b)]. For n-bit full adders ($n \geq 1$), there were $50n-28$ transistors in a classical circuit [Fig. 6(c)], whereas there were $44n-28$ in a fusion-gate circuit [Fig. 6(d)]. Again, the fusion-gate circuit has a little advantage in the number of transistors, but the difference will increase as n increases.

The comparison of the number of transistors between classical and fusion-gate logic is summarized in Fig. 7. Except for an AND logic ($n \geq 4$), the number of transistors in fusion-gate logic is always smaller than that of transistors in classical logic circuits, especially in majority logic gates.

The computational ability of the proposed methodology can be evaluated by calculating logical functions of a 2D array of fusion gates at each output node. The 2D computing matrix, where x_i and y_i represent the horizontal and vertical inputs and X_i and Y_i the horizontal and vertical outputs, is shown in Fig. 8. For simplicity, here we assume y_i = logical '1'. Then, the horizontal and vertical outputs are represented by

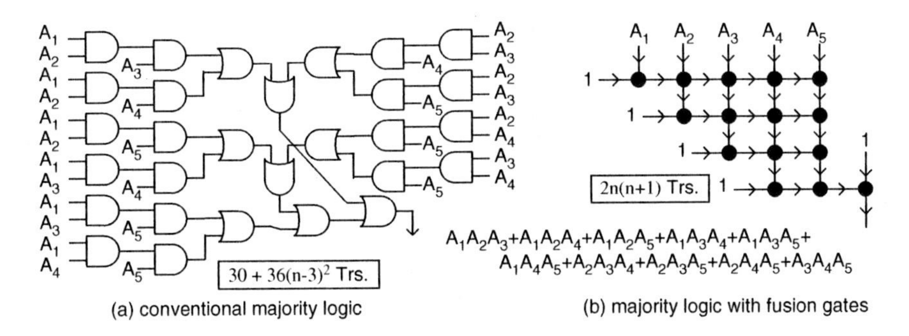

Fig. 5. Classical and fusion-gate logic architectures of majority logic circuits.

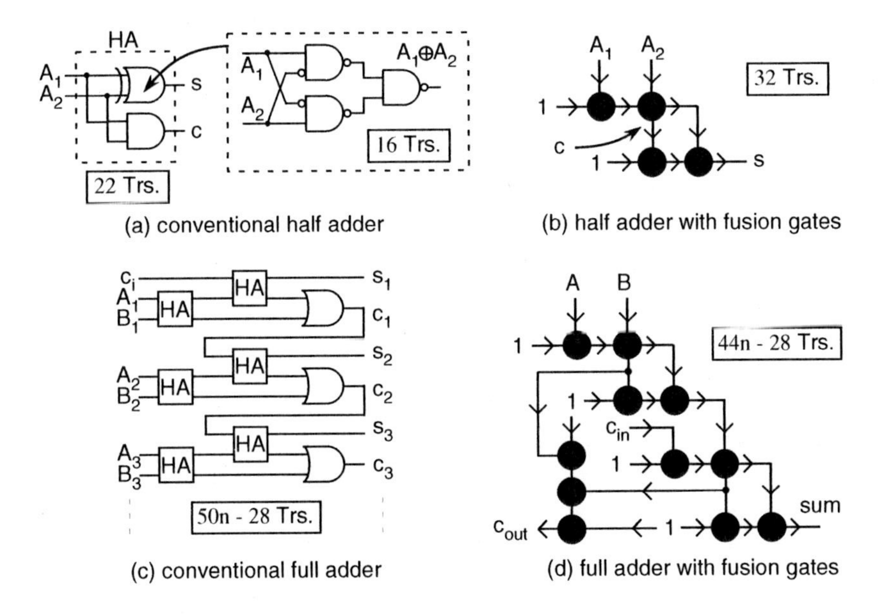

Fig. 6. Classical and fusion-gate logic architectures of half- and full-adder circuits.

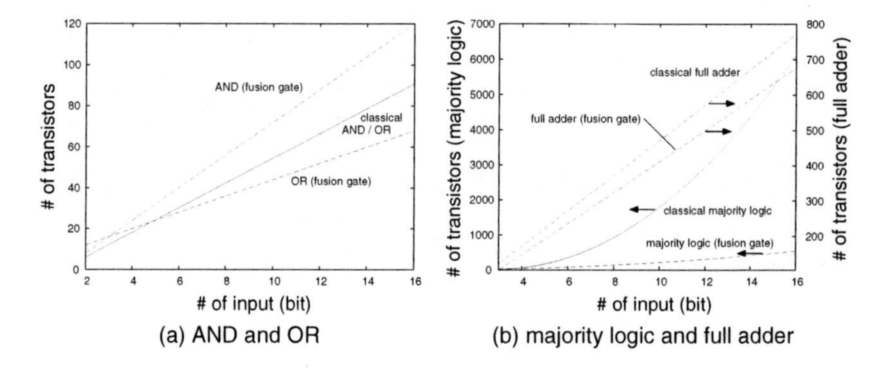

Fig. 7. Comparison of the number of transistors between classical and fusion-gate logic.

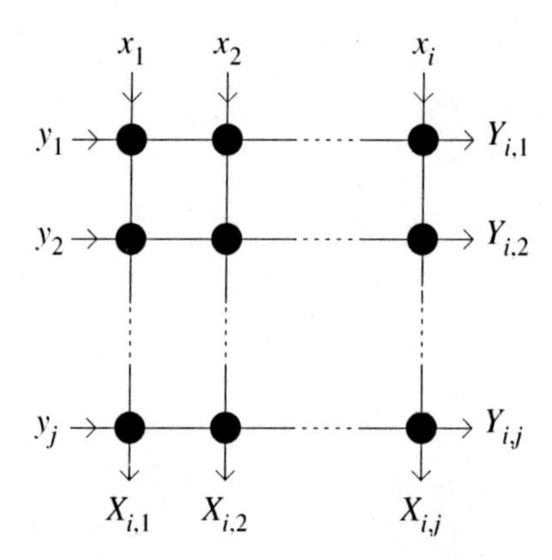

Fig. 8. Fusion-gate array of fusion gates on 2D rectangular grid where x_i and y_i represent the horizontal and vertical inputs and X_i and Y_i the horizontal and vertical outputs.

the following difference equations:

$$Y_{i,1} = \overline{\sum_{k=1}^{i} X_k}, \quad (i \geq 1) \tag{1}$$

$$Y_{i,j} = \overline{\sum_{k=1}^{i-(j-1)} Y_{i,j-1}}, \quad (i \geq j) \tag{2}$$

$$X_{i,1} = X_i \sum_{k=1}^{i-1} X_k, \quad (i \geq 2) \tag{3}$$

$$X_{i,j} = X_i \sum_{k=1}^{i-j} X_{i,j-1}, \quad (i \geq j). \tag{4}$$

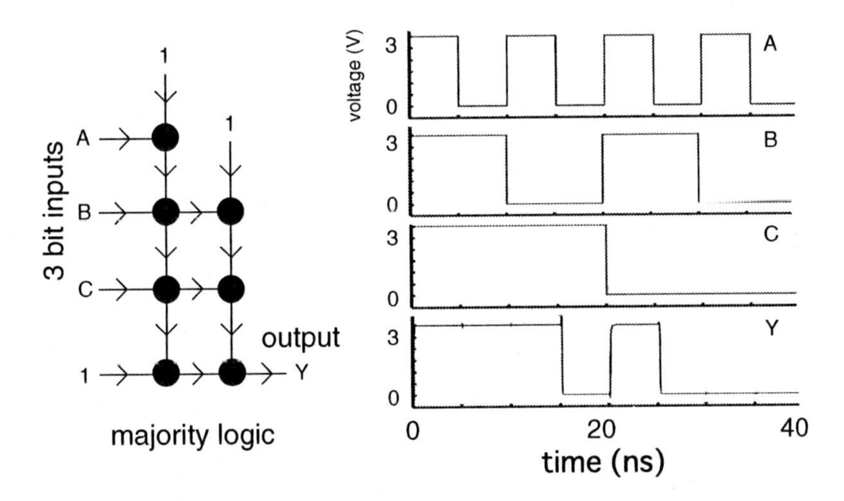

Fig. 9. Example construction of majority logic gate and its simulation results with 0.35-μm digital CMOS parameters (MOSIS, Vendor TSMC).

An example of implementation and simulations of three-input majority logic gates is shown in Fig. 9. Seven fusion gates, i.e., 28 transistors for high-frequency operation, are required for this function, whereas in conventional architecture, 30 transistors are necessary for the same function. Using the fusion-gate circuit shown in Fig. 3(b), we simulated the majority logic circuit by using SPICE with 0.35-μm digital CMOS parameters (MOSIS, Vendor TSMC, with minimum-sized transistors). The results obtained where A, B, and C represent the input, while Y represents the output are shown in Fig. 9 (right). The clock frequency was 100 MHz. The rise time of the output was 0.3 ns for this parameter set. The operation speed is thus significantly faster than that of the two-transistor circuit.

Let us consider the number of transistors per function shown in Figs. 2(c)-(e). A two-transistor fusion gate [Fig. 3(a)] is used for the time being. In the case of NOT, two transistors are necessary; that is the same number of transistors as that of a conventional inverter circuit. For AND and NOR functions, four transistors are required, which is half the number of transistors in a combinational circuit of conventional AND and NOR circuits. In the case of AND and OR, six transistors are required, whereas ten are required in conventional circuits. Therefore, for low clock-frequency applications, fusion-gate logical computing with two-transistor fusion gates certainly decreases the number of transistors in the circuit network. For high-speed applications, four-transistor fusion gates have to be used; however, the number of transistors is doubled in this case.

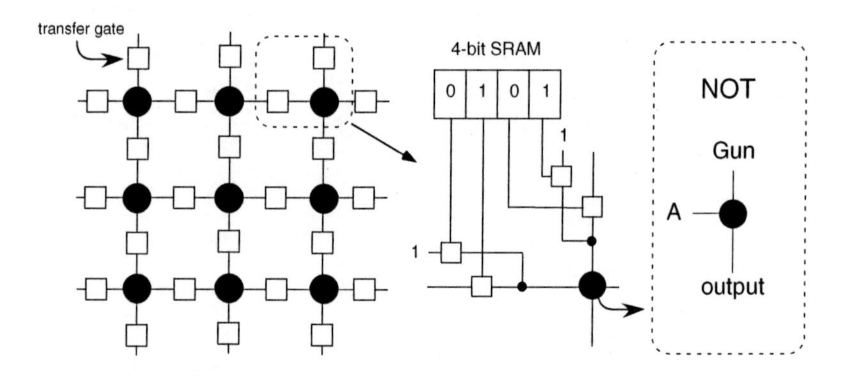

Fig. 10. Possible construction of reconfigurable logic architecture.

What are the merits of fusion-gate architecture for high-speed applications? There are two types of answers: introducing yet another device and completely new functions. In the former case, a single-electron reaction-diffusion device [4] is a possible candidate. For the latter case, let us consider 'reconfigurable' functions. A basic idea of the reconfigurable logic architecture using fusion gates is shown in Fig. 10. Each fusion gate is regularly arranged on a 2D chip surface and is locally connected to other fusion gates via transfer gates. In this construction a new unit gate consists of four transfer gates, a single fusion gate circuit, and a four-bit memory circuit in which two bits give static inputs to the gate, and the remainder is for selecting the signal flow. As an example, a NOT function is depicted in the figure.

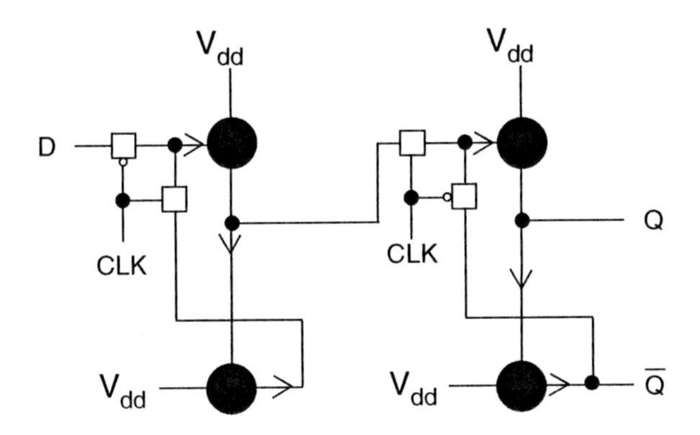

Fig. 11. Construction of D-type flip flop for sequential circuits.

Arbitrary combinational circuits for arithmetic modules can be constructed by the proposed fusion-gate circuits, but how about sequential circuits for practical computation? In other words, how can we implement static memory, e.g., flip-flop, circuits? A possible answer to this question is shown in Fig. 11. Remember that the fusion gate circuit shown in Fig. 3(b) consists of two inverters. Therefore, by rewiring the fusion gate circuit, one can construct a D-type flip-flop (D-FF) circuit, as shown in Fig. 11. Although four additional transfer-gates are required, D-FF circuits can be constructed while maintaining the basic construction of a 2D array of fusion gate circuits.

4 Summary

We described a method of designing logic circuits with fusion gates which is inspired by collision-based reaction-diffusion computing. First, we introduced a new interpretation of collision-based computing, especially concerning a limited direction of wave fragments and infinite transition speed. This simplified construction of the computing media significantly. Second, we showed that basic logical functions could be represented in terms of our unit operator that calculated both $\overline{A}B$ and $A\overline{B}$ for inputs A and B. Third, two basic MOS circuits were introduced; one consisted of two transistors but operated very slowly, whereas the other consisted of four transistors but operated much faster than the two-transistor circuit.

In the constructions of basic logic functions, the number of transistors in the two-transistor circuit was smaller than that of the corresponding conventional circuits. However, the number in the four-transistor circuit was larger than them. The combination of the fusion gates produces multiple functions, e.g., an AND circuit can compute NOR simultaneously, so we should build optimization theories for generating multiple-input arbitrary functions. Finally, we introduced a possible construction of D-type flip-flop circuits for constructing sequential circuits.

Acknowledgments

The authors wish to thank Professor Andrew Adamatzky of the University of the West of England for valuable discussions and suggestions during the research and Professor Masayuki Ikebe of Hokkaido University for suggestions concerning various CMOS circuits. This study was supported in part by the Industrial Technology Research Grant Program in 2004 from the New Energy and Industrial Technology Development Organization (NEDO) of Japan and by a grant-in-aid for young scientists [(B)17760269] from the Ministry of Education, Culture, Sports, Science, and Technology (MEXT) of Japan.

References

1. A. Adamatzky, Universal dynamical computation in multi-dimensional excitable lattices, *Int. J. Theor. Phys.*, vol. 37, pp. 3069–3108, 1998.
2. A. Adamatzky, Editor, *Collision-Based Computing*, Springer-Verlag, 2002.
3. A. Adamatzky, Computing with waves in chemical media: Massively parallel reaction-diffusion processors, *IEICE Trans. Electron.*, vol. E87-C, no. 11, pp. 1748–1756, 2004.
4. A. Adamatzky A., B. De Lacy Costello, and T. Asai, *Reaction-Diffusion Computers*. Elsevier, UK, 2005.
5. M. Alioto and G. Palumbo, *Model and Design of Bipolar and MOS Current-Mode Logic: CML, ECL and SCL Digital Circuits*, Springer, 2005.
6. N. Asahi, M. Akazawa, and Y. Amemiya, Single-electron logic systems based on the binary decision diagram, *IEICE Trans. Electronics*, vol. E81-C, no. 1, pp. 49–56, 1998.
7. E. R. Berlekamp, J. H. Conway, and R. L. Guy, *Winning Ways for your Mathematical Plays*. Vol. 2. Academic Press, 1982.
8. F. Fredkin and T. Toffoli, Conservative logic, *Int. J. Theor. Phys.*, vol. 21, pp. 219–253, 1982.
9. N. Margolus, Physics-like models of computation, *Physica D*, vol. 10, pp. 81–95, 1984.

10. Y. Matsubara, T. Asai, T. Hirose, and Y. Amemiya, Reaction-diffusion chip implementing excitable lattices with multiple-valued cellular automata, *IEICE ELEX*, vol. 1, no. 9, pp. 248–252, 2004.
11. I. N. Motoike and K. Yoshikawa, Information operations with multiple pulses on an excitable field, *Chaos, Solitons and Fractals*, vol. 17, pp. 455–461, 2003.
12. I. Motoike and K. Yoshikawa, Information operations with an excitable field, *Phy. Rev. E*, vol. 59, no. 5, pp. 5354–5360, 1999.
13. D. E. Muller, Application of Boolean Algebra to Switching Circuit Design and to Error Detection", *IRE Trans. on Electr. Comp.*, vol. EC-3, pp. 6–12, 1954.
14. I. S. Reed, A Class of Multiple-Error-Correcting Codes and Their Decoding Scheme", *IRE Trans. on Inform. Th.*, vol. PGIT-4, pp. 38–49, 1954.
15. R. S. Shelar and S. S. Sapatnekar, BDD decomposition for the synthesis of high performance PTL circuits, *Workshop Notes of IEEE IWLS 2001*, pp. 298–303.
16. H. Soeleman and K. Roy, Ultra-low power digital subthreshold logic circuits, in *Proc. 1999 Int. Symp. on low power electronics and design*, pp. 94–96.
17. H. Soeleman, K. Roy, and B. C. Paul, Robust subthreshold logic for ultra-low power operation, *IEEE Trans. on Very Large Scale Integration (VLSI) Systems*, vol. 9, no. 1, pp. 90–99, 2001.
18. M. Song and K. Asada, Design of low power digital VLSI circuits based on a novel pass-transistor logic, *IEICE Trans. Electronics*, vol. E81-C, no. 11, pp. 1740–1749, 1998.
19. K. Steiglitz, I. Kamal, and A. Watson, Embedded computation in one–dimensional automata by phase coding solitons, *IEEE Trans. Comp.*, vol. 37, pp. 138–145, 1988.
20. T. Yamada, Y. Kinoshita, S. Kasai, H. Hasegawa, and Y. Amemiya, Quantum-dot logic circuits based on the shared binary decision diagram, *Jpn. J. Appl. Phys.*, vol. 40, no. 7, pp. 4485–4488, 2001.

Towards machine learning control of chemical computers

Adam Budd[1], Christopher Stone[1], Jonathan Masere[2], Andrew Adamatzky[1], Ben De Lacy Costello[2], and Larry Bull[1]

[1] Faculty of Computing, Engineering & Mathematics, University of the West of England
[2] Faculty of Applied Sciences, University of the West of England
Coldharbour Lane, Frenchay
Britsol BS16 1QY, U.K.
Adam3-Budd@uwe.ac.uk

Abstract. The behaviour of pulses of Belousov-Zhabotinsky (BZ) reaction-diffusion waves can be controlled automatically through machine learning. By extension, a form of chemical network computing, i.e., a massively parallel non-linear computer, can be realised by such an approach. In this initial study a light-sensitive sub-excitable BZ reaction in which a checkerboard image comprising of varying light intensity cells is projected onto the surface of a thin silica gel impregnated with tris(bipyridyl) ruthenium (II) catalyst and indicator is used to make the network. As a catalyst BZ solution is swept past the gel, pulses of wave fragments are injected into the checkerboard grid resulting in rich spatio-temporal behaviour. This behaviour is shown experimentally to be repeatable under the same light projections. A machine learning approach, a learning classifier system, is then shown able to direct the fragments to an arbitrary position through dynamic control of the light intensity within each cell in both simulated and real chemical systems.

1 Introduction

There is growing interest in research into the development of hybrid wetware-silicon devices focused on exploiting their potential for non-linear computing [1, 3]. The aim is to harness the as yet only partially understood intricate dynamics of non-linear media to perform complex computations (potentially) more effectively than with traditional architectures and to further the understanding of how such systems function. The area provides the prospect of radically new forms of machines and is enabled by improving capabilities in wetware-silicon interfacing. We are developing an approach by which networks of non-linear media – reaction-diffusion systems — can be produced to achieve a user-defined computation in a way that allows control of the media used. Machine Learning algorithms are used to design the appropriate network structures by searching a defined behavioural space to create a computing resource capable of satisfying a given objective(s). In this paper we examine the underlying dynamics of the chosen Belousov-Zhabotinsky (BZ) [43] reaction-diffusion system in which the networks are created via light and present initial results from the general control/programming methodology.

Excitable and oscillating chemical systems have been used to solve a number of computational tasks [1, 3], such as implementing logical circuits [35, 39], image processing [28], shortest path problems [36] and memory [32]. In addition chemical diodes [4], coincidence detectors [16] and transformers where a periodic input signal of waves may be modulated by the barrier into a complex output signal depending on the gap width and frequency of the input [34] have all been demonstrated experimentally. However, to some degree the lack of compartmentalisation in these simple chemical systems limits the domain of solvable tasks thus making it difficult to realise general-purpose computing. This proposed methodology of utilising networks of coupled oscillating chemical reactions may provide a solution. The fact that these coupled oscillators can be controlled via the application of external fields such as light provides a possible method for undertaking a number of complex computations provided an effective methodology for realising large scale networks can be found.

A number of experimental and theoretical constructs utilising networks of chemical reactions to implement computation have been described. These chemical systems act as simple models for networks of coupled oscillators such as neurons, circadian pacemakers and other biological systems [26]. Over 30 years ago the construction of logic gates in a bistable chemical system was described by Rossler [33]. Ross and co-workers [19] produced a theoretical construct suggesting the use of "chemical" reactor systems coupled by mass flow for implementing logic gates neural networks and finite-state machines. In further work Hjelmfelt et al [17, 18] simulated a pattern recognition device constructed from large networks of mass-coupled chemical reactors containing a bistable iodate-arsenous acid reaction. They encoded arbitrary patterns of low and high iodide concentrations in the network of 36 coupled reactors. When the network is initialized with a pattern similar to the encoded one then errors in the initial pattern are corrected bringing about the regeneration of the stored pattern. However, if the pattern is not similar then the network evolves to a homogenous state signalling non-recognition.

In related experimental work [29] used a network of eight bistable mass coupled chemical reactors (via 16 tubes) to implement pattern recognition operations. They demonstrated experimentally that stored patterns of high and low iodine concentrations could be recalled (stable output state) if similar patterns were used as input data to the programmed network. This highlights how a programmable parallel processor could be constructed from coupled chemical reactors. This described chemical system has many properties similar to parallel neural networks. In other work [30] described methods of constructing logical gates using a series of flow rate coupled continuous flow stirred tank reactors (CSTR) containing a bistable nonlinear chemical reaction. The minimal bromate reaction involves the oxidation of cerium(III) (Ce^{3+}) ions by bromate in the presence of bromide and sulphuric acid. In the reaction the Ce^{4+} concentration state is considered as ("0" "false") and ("1" "true") if a given steady state is within 10% of the minimal (maximal) value. The reactors were flow rate coupled according to rules given by a feedforward neural network run using a PC. The experiment is started by feeding in two "true" states to the input reactors and then switching the flow rates to generate "true"–"false", "false"–"true" and "false"–"false". In this three coupled reactor system the AND (output "true" if inputs are both high Ce^{4+}, "true"), OR (output "true" if one of the inputs is "true"), NAND (output "true" if one of the inputs is "false") and NOR gates (output "true" if both of the inputs are "false") could be realised. However to construct XOR and XNOR gates two additional reactors (a hidden layer) were required. These composite gates are solved by interlinking AND and OR gates and their negations. In their work coupling was implemented by computer but they suggested that true chemical computing of some Boolean functions may be achieved by using the outflows of reactors as the inflows to other reactors i.e. serial mass coupling. As yet no large scale experimental network implementations have been undertaken mainly due to the complexity of analysing and controlling so many reactors. That said there have been many experimental studies carried out involving coupled oscillating and bistable systems [5,6,11,12,23,37]. The reactions are coupled together either physically by diffusion or an electrical connection or chemically, by having two oscillators that share a common chemical species. The effects observed include multistability, synchronisation, in-phase and out of phase entrainment, amplitude or "oscillator death", the cessation of oscillation in two coupled oscillating systems, or the converse, "rythmogenesis", in which coupling two systems at steady state causes them to start oscillating [13].

In this paper we adapt a system described by [40] and explore the computational potential based on the movement and control of wave fragments. In the system they describe the application of Gaussian noise (where the mean light level is fixed at the subexcitable threshold of the reaction) in the form of light projected onto a thin layer of the light sensitive analogue of the BZ reaction was observed to induce wave formation and subsequently "avalanche behaviour" whereby a proliferation of open ended excitation wave fragments were formed. Interestingly calcium waves induced in networks of cultured glial cells [24] display similar features to the ones identified in this chemical system which the authors postulated may provide a possible mechanism for long-range signalling and memory in neuronal tissues.

Machine Learning techniques, such as Evolutionary Algorithms [31] (EAs) and Reinforcement Learning (RL) [38], are being increasingly used in the design of complex systems. Example applications include data mining, time series analysis, scheduling, process control, robotics and electronic circuit design. Such techniques can be used for the design of computational resources in a way that offers substantial promise for application in non-linear media computing since the algorithms are almost independent of the medium in which the computation occurs. This is important in order to achieve effective non-linear media computing since they do not need to directly manipulate the material to facilitate learning and the task itself can be defined in a fairly unsupervised manner. In contrast, most traditional learning algorithms use techniques that require detailed knowledge of and control over the computing substrate involved. In this paper we control the BZ network via a reinforcement learning approach which uses evolutionary computing to create generalizations over the state-action space – Hollands Learning Classifier System [22], in particular a form known as XCS [42].

The paper is arranged as follows: The next section describes the subexcitable BZ system which forms the basis of this chemical computing research and results from investigations into its basic properties. The next section describes a computational model of the system and the chosen machine learning approach. Initial results from using the machine learner to control the simulated and real chemical system are then presented.

2 Experimental system

2.1 Material and equipment

Sodium bromate, sodium bromide, malonic acid, sulphuric acid, tris (bipyridyl) ruthenium (II) chloride, 27% sodium silicate solution stabilized in 4.9 M sodium hydroxide were purchased from Aldrich and used as received unless stated otherwise.

An InFocus Model Projector was used to illuminate the computer-controlled image. Images were captured using a Panasonic NV-GS11 digital video camera. The microscope slide was immersed in the continuously fed reaction solution contained in a custom-designed Petri dish, designed by Radleys, with a water jacket thermostatted at 22 °C. A Watson Marlow 205U multi-channel peristaltic pump was used to pump the reaction solution into the reactor and remove the effluent.

3 Experimental procedures

3.1 Making gels

A stock solution of the sodium silicate solution was prepared by mixing 222 mL of the purchased sodium silicate solution with 57 mL of 2 M sulphuric acid and 187 mL of deionised water, similar to the procedure used by [40]. Pre-cured solutions for making gels were prepared by mixing 5 mL of the acidified silicate solution with a solution consisting of 1.3 mL of 1.0 M sulphuric acid and 1.2 mL of 0.025 M tris(bipyridyl) ruthenium (II) chloride. Using capillary action, portions of this solution were transferred onto microscope slides with 100 μm shims and Plexiglas covers. The transferred solutions were left for 3 hours to permit complete gellation after which the covers and shims were removed and the gels washed in deionised water to remove residual tris(bipyridyl) ruthenium (II) chloride and the sodium chloride byproduct. The gels were 26 mm by 26 mm, with a wet thickness of approximately 100 μm. The gels were stored under water and rinsed right before use.

3.2 Catalyst-free reaction mixture

The bulk of the catalyst-free reaction mixture was freshly prepared in 300 mL batches, which involved the in situ synthesis of stoichiometric bromomalonic acid from malonic acid and bromine generated from the partial reduction of sodium bromate. The catalyst-free reaction solution consisted of the 0.36 M sodium bromate, 0.0825 M malonic acid, 0.18 M sulphuric acid and 0.165 M bromomalonic acid. To minimize the deterioration during the experiment, this solution was kept in an ice bath. This solution was continuously fed into the thermostatted reactor, with a reactor residence time of 30 minutes.

3.3 Experimental setup

The spatially distributed excitable field on the surface of the gel was made possible by the projection of a 10-by-10 cell checkerboard grid pattern generated using a computer. After [40], the checkerboard image comprised the heterogeneous network of Gaussian distributed light levels, with a mean at the sub-excitable threshold, from a low of 0.394 mW cm^{-2} to a high of 9.97 mW cm^{-2} intensity cells in 256 bins, representing excitable and non-excitable domains respectively. A digital video camera was used to capture the chemical wave fragments. A diagrammatic representation of the experimental setup is shown in Fig. 1.

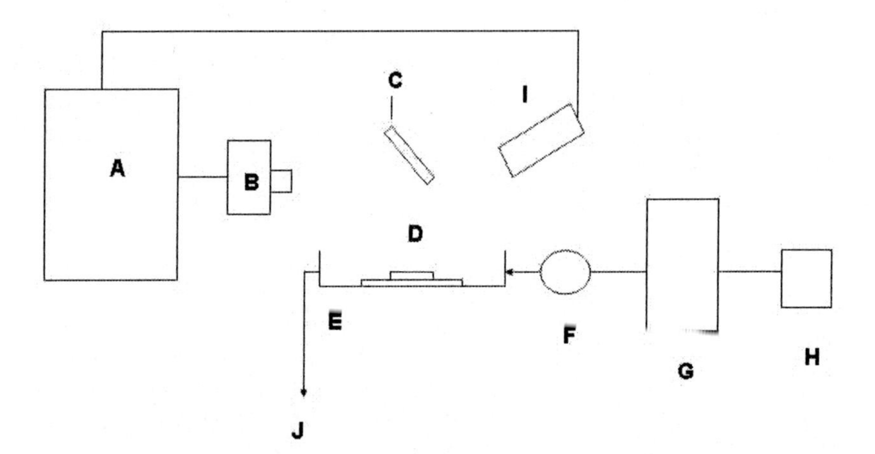

Fig. 1. A block diagram of the experimental setup where the computer, projector, mirror, microscope slide with the catalyst-laden gel, thermostatted Petri dish, peristaltic pump, thermostatted water bath, reservoir of catalyst-free reaction solution, digital camcorder and, the effluent flow are designated by A, B, C, D, E, F. G, H, I and J, respectively. The catalyst-free reaction solution reservoir was kept in an ice bath during the experiment.

3.4 Data capturing and image processing

A checkerboard grid pattern was projected onto the catalyst-laden gel through a 455 nm narrow bandpass interference filter and 100/100 mm focal length lens pair and mirror assembly. The size of the projected grid was approximately 20mm square. Every 10 seconds, the checkerboard pattern was replaced with a uniform grey level of 9.97 mW cm^{-2} for 400 ms during which time an image of the BZ waves on the gel was captured. The purpose of removing the grid pattern during this period was to allow activity on the gel to be more visible to the camera and assist in subsequent image processing of chemical activity.

Captured images were processed to identify chemical wave activity. This was done by differencing successive images on a pixel by pixel basis to create a black and white thresholded image. Each pixel in the black and white image was set to white, corresponding to chemical activity; if the intensity of the red or blue channels differed in successive images by more than 1.95%. Pixels at locations not meeting this criterion were set to black. The thresholded images were automatically despeckled and manually edited to remove artefacts of the experiment, such as glare from the projector and bubbles from the oxidative decarboxylation of malonic acid and bromomalonic acid. The images were cropped to the grid location and the grid superimposed on the thresholded images to aid analysis of the results.

3.5 Experimental system behaviour

Computationally, the described system has 100 inputs and potentially as many outputs. In this study we have examined the use of single pulses of excitation into the bottom of the grid. Each pulse expands and grows as a fragmented front while traversing the network of excitable domains of varying light intensity. Multiple disproportionate collisions result in daughter fragments, mutual fragment extinctions, etc. (Figs. 2 and 3). That is, oscillatory/clustering behaviour can be observed -- wave fragments and their collisions equate to information transfer [1,2] – the fragment dynamics are mediated by local light levels, collisions, boundaries, trajectories of previous fragments, diffusion processes in adjacent cells and beyond, etc.

The spatio-temporal evolution of the system is affected by the series of light levels projected into each cell. To be useful as a computational formalism, the behaviour of the system under a given light "program", i.e., a series of levels in each given cell, must be sufficiently repeatable. That is, it must be possible to exert control at some appropriate level over the evolution and dynamics of the excitable wave fragments. Figures 2 and 3 show example results from three random light programs and three runs using the same light programs where we aimed to control the evolution of the fragmentation pattern from a single input, thereby enabling us to assess the repeatability/stability of the chemical system. As can be seen, in the former case (Fig. 2) there is no obvious correspondence in the spatio-temporal behaviour whereas the behaviour under the same program is remarkably similar (Fig. 3).

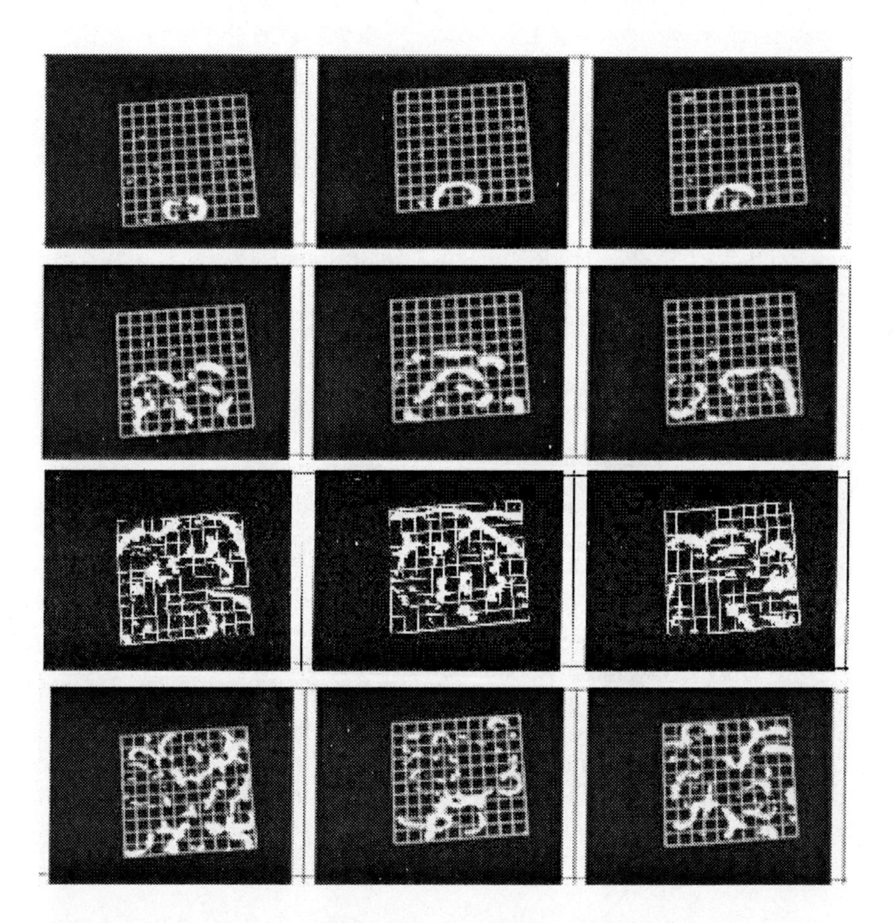

Fig. 2. Example spatio-temporal evolution of the chemical system under different light programs, with slides taken at 10 second intervals.

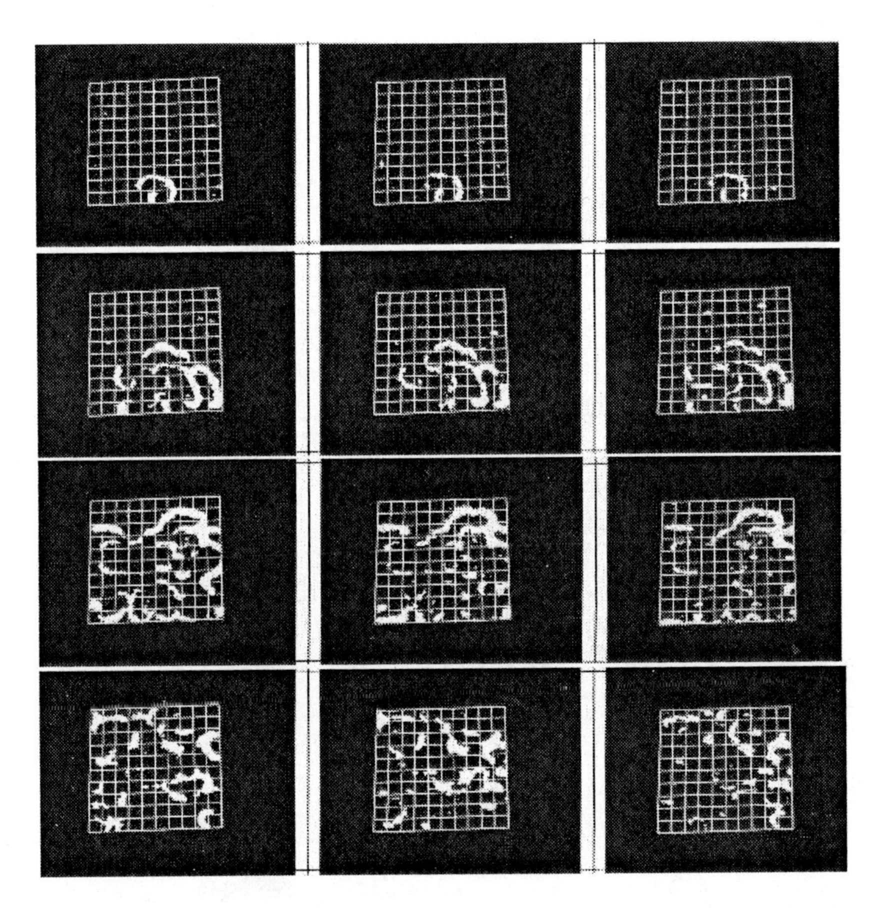

Fig. 3. Example spatio-temporal evolution of the chemical system under the same light program, with slides taken at 10 second intervals.

4 Computational system

Model The features of this system were simulated using a two-variable Oregonator model modified to account for the photochemistry [15, 27]:

$$\frac{\partial u}{\partial t} = \frac{1}{\varepsilon}\left\{u - u^2 - (fv + \Phi)\frac{u-q}{u+q}\right\} + D_u \nabla^2 u$$
$$\frac{\partial v}{\partial t} = u - v \quad (1)$$

The variables u and v in equation 1 represent the instantaneous local concentrations of the bromous acid autocatalyst and the oxidized form of the catalyst, $HBrO_2$ and tris (bipyridyl) Ru (III), respectively, scaled to dimensionless quantities. The ratio of the time scales of the two variables, u and v, is represented by ε, which depends on the rate constants and reagent concentration; f is a stoichiometric coefficient. The rate of the photo-induced bromide production is designated by Φ, which also denotes the excitability of the system in which low light intensities facilitate excitation while high intensities result in the production of bromide that inhibits the process, experimentally verified by [25]. The scaling parameter, q, depends on reaction rates only. The system was integrated using the Euler method with a five-node Laplacian operator, time step Δt=0.001 and grid point spacing Δx=0.15. The diffusion coefficient, Du, of species u was unity, while that of species v was set to zero as the catalyst was immobilized in gel.

4.1 Machine learning algorithm

XCS is a relatively recent development of Hollands Learning Classifier System formalism and has been shown able to tackle many complex tasks effectively (see [8] for examples). It consists of a limited size population [P] of classifiers (rules). Each classifier is in the form of "IF condition THEN action" (*condition* → *action*) and has a number of associated parameters. Conditions traditionally consist of a trinary representation, {0,1,#}, where the wildcard symbol facilitates generalization, and actions are binary strings.

On each time step a match set [M] is created. A system prediction is then formed for each action in [M] according to a fitness-weighted average of the predictions of rules in each action set [A]. The system action is then selected either deterministically or stochastically based on the fitness-weighted predictions (usually 0.5 probability per trial). If [M] is empty a covering heuristic is used which creates a random condition to match the given input and then assigns it to a rule for each possible action.

Fitness reinforcement in XCS consists of updating three parameters, ε, p and F for each appropriate rule; the fitness is updated according to the relative accuracy of the rule within the set in five steps:

1. Each rules error is updated: $\varepsilon_j = \varepsilon_j + \beta(|P - p_j| - \varepsilon_j)$ where is a learning rate constant.
2. Rule predictions are then updated: $p_j = p_j + \beta(P\text{-}p_j)$
3. Each rules accuracy κ_j is determined: $\kappa_j = \alpha(\varepsilon_0/\varepsilon)^v$ or κ=1 where $\varepsilon < \varepsilon_0$ v,α and ε_0 are constants controlling the shape of the accuracy function.
4. A relative accuracy κ_j' is determined for each rule by dividing its accuracy by the total of the accuracies in the action set.
5. The relative accuracy is then used to adjust the classifiers fitness F_j using the moyenne adaptive modifee (MAM) procedure: If the fitness has been adjusted $1/\beta$ times, $F_j = F_j + \beta(\kappa_j\text{'} - F_j)$. Otherwise F_j is set to the average of the values of κ' seen so far.

In short, in XCS fitness is an inverse function of the error in reward prediction, with errors below ε_0 not reducing fitness. The maximum *P(ai)* of the systems prediction array is discounted by a factor γ and used to update rules from the previous time step and an external reward may be received from the environment. Thus XCS exploits a form of Q-learning [41] in its reinforcement procedure.

A Genetic Algorithm (GA) [21] acts in action sets [A], i.e., niches. Two rules are selected based on fitness from within the chosen [A]. Two-point crossover is applied at rate χ and point mutations at rate μ. Rule replacement is global and based on the estimated size of each action set a rule participates in with the aim of balancing resources across niches. The GA is triggered within a given action set based on the average time since the members of the niche last participated in a GA (after [7]).

The intention in XCS is to form a complete and accurate mapping of the problem space through efficient generalizations. In reinforcement learning terms, XCS learns a value function over the complete state/action space. In this way, XCS represents a means of using temporal difference learning on complex problems where the number of possible state-action combinations is very large (other approaches have been suggested, such a neural networks – see [38] for an overview). The reader is referred to [10] for an algorithmic description of XCS and [9] for an overview of current formal understanding of its operations.

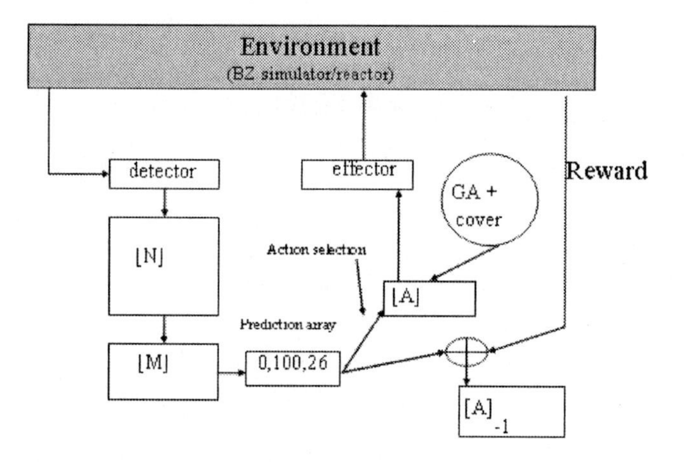

Fig. 4. Schematic of the XCS Learning Classifier System.

4.2 XCS control: simulator

The aforementioned model of the BZ system has been interfaced to an implementation of XCS in a way which approximates the envisaged hardware-wetware scenario. A 3x3 grid is initialized with a pulse of excitation in the bottom middle cell as in the wetware experiments described in section 2. Two light levels have been used thus far: one is sufficiently high to inhibit the reaction; and, the other low enough to enable it. The modeled chemical system is then simulated for 10 seconds of real time. A 9-bit binary description of the 3x3 grid is then passed to the XCS. Each bit corresponds to a cell and it is set to true if the average level of activity within the given cell is greater than a pre-determined threshold. The XCS returns a 9-bit action string, each bit of which indicates whether the high (discovered by experimentation Φ=0.197932817) or low (discovered by experimentation Φ=0.007791641) intensity light should be projected onto the given cell. Another 10 seconds of real time are then simulated, etc. until either a maximum number of iterations has passed or the emergent spatial-temporal dynamics of the system match a required configuration. In this initial work, a fragment is required to exist in the middle left-hand cell of the grid only. At such a time, a numerical reward of 1000 is given the system and the system reset for another learning trial. To be able to obtain this behavior reliably, it has been found beneficial to use an intermediate reward of 500 in the presence of a fragment in the target cell, regardless of the activity on the rest of the grid (see [14] for related discussions). The XCS parameters used for this were (largely based on [42]): N=30.000, β=0.2, μ=0.04, χ=0.8, γ=0.71, θ_{del}=20, δ=0.1, ε_0=10, α=0.1, ν=5.0, θ_{mna}=512, θ_{GA}=25, $\varepsilon_I = \varepsilon_I = F_I$=10.0, $p_{\#}$=0.33. Other parameters for the BZ model were ε=0.022, f=1.4, q=0.002.

Figure 5 shows a typical light program and associated wave fragment behaviour sequence, here taking 9 steps to solve the problem, which appears optimal with the given parameters of the simulator and allowed time between XCS control steps. Figure 6 shows the reward received by the learner per trial, averaged over 3 runs. As can be seen, this approaches the maximum of 1000 in the time considered; the XCS controller is reliable in its ability to develop a fragment controller in the given scenario.

4.3 Control chemistry

Given the success of the simulation experiments, the machine learning system described above was connected to the chemical system described in section 2. The scenario for the chemical experiments was the same as for the simulations, although it must be noted that there is a slightly longer delay before the XCS is able to control the initial light levels once the pulse is added to the grid due to the image processing required. Figure 7 shows an example result using the same parameters as before. This solution was discovered on trial 4 and then refined over the subsequent three trials to that shown. It can be seen that XCS has learned to control the fragment in the real chemical system as it did in the simulation.

5 Conclusions

Excitable and oscillating chemical systems have previously been used to solve a number of computational tasks. However we suggest that the lack of compartmentalisation in the majority of the previous systems limits the domain of solvable tasks thus making it difficult to realise general-purpose computing. We propose that utilising networks of coupled oscillating chemical reactions will open chemical computing to wider domains. In this paper we have presented initial results from a methodology by which to achieve the complex task of designing such systems — through the use of machine learning techniques. We have shown using both simulated and real systems that it is possible control the behaviour of a light-sensitive BZ reaction using XCS.

Current work is extending the scenario presented to include larger grids containing many more concurrent fragments of excitation.

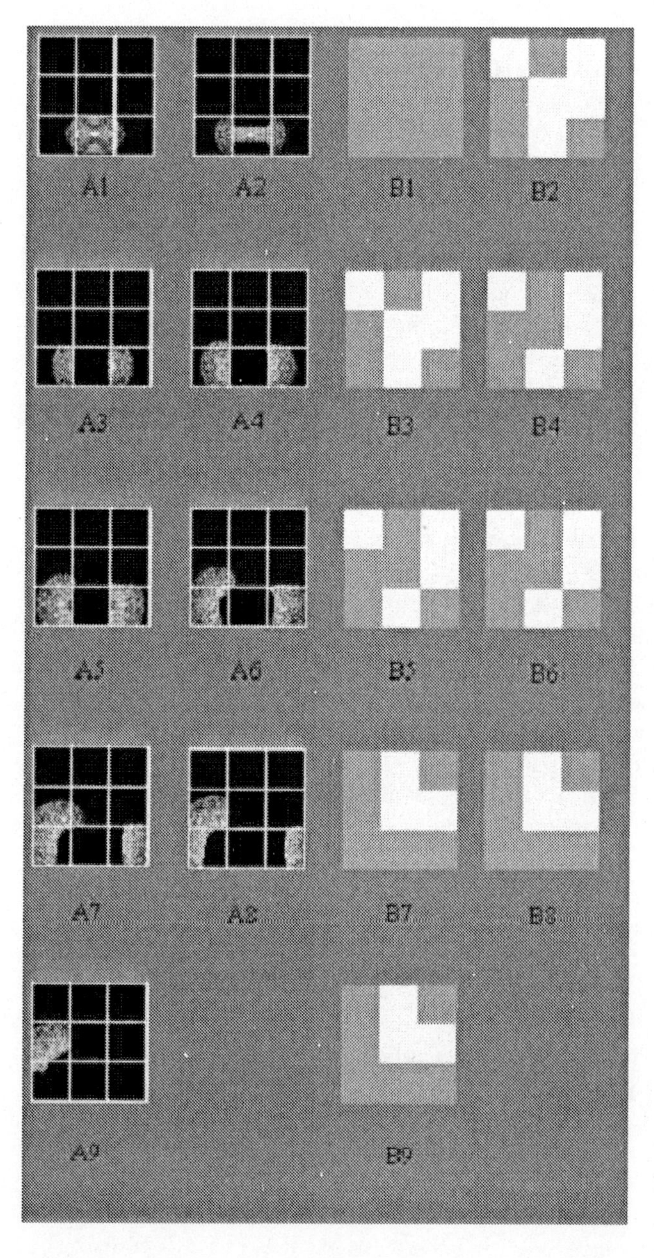

Fig. 5. Example control of the simulated chemical system (A) under the learned light programs (B).

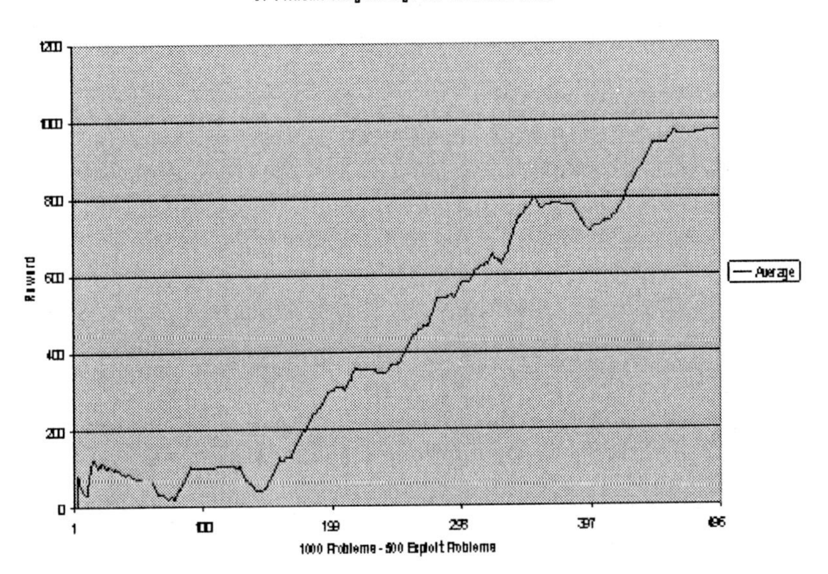

Fig. 6. Showing the average reward received by the XCS controllers with increasing number of learning cycles or problems.

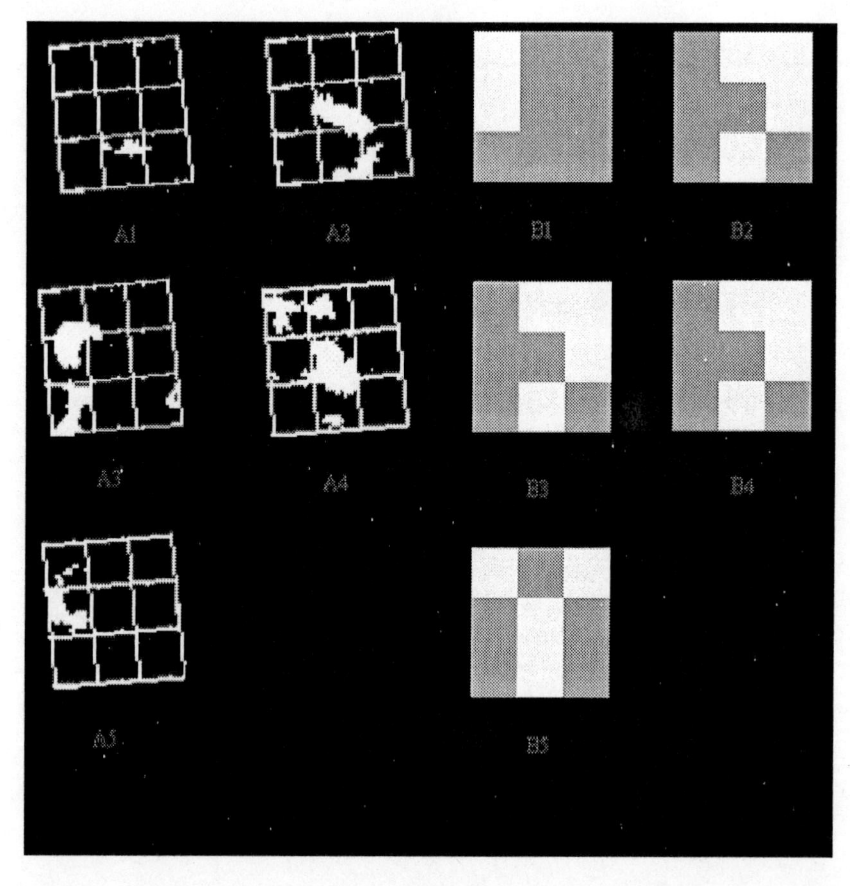

Fig. 7. Example control of the real chemical system (A) under the learned light programs (B). Skewing of cell images is due to camera set-up.

6 Acknowledgements

This work was supported by EPSRC grant no. GR/T11029/01. Thanks to Annette Taylor and Rita Toth for useful discussions during the course of this work.

References

1. Adamatzky, A. (2001) *Computation in Nonlinear Media and Automata Collectives.* IoP Publishing.
2. Adamatzky, A., (2003)(ed) *Collision Based Computing.* Springer.
3. Adamatzky, A., De Lacy Costello, B., & Asai, T. (2005) *Reaction-Diffusion Computers.* Elsevier
4. Agladze K, Aliev RR, Yamaguhi T & Yoshikawa K. (1996) Chemical Diode. *J Phys Chem.* 100: 13895–13897.
5. Bar-Eli, K. & Reuveni, S. (1985) Stable stationary-states of coupled chemical oscillators. Experimental evidence, *J Phys Chem.* 89: 1329–1330.
6. Bar-Eli, K. (1985) On the stability of coupled oscillators. *Physica D* 14: 242–252.
7. Booker, L. (1989) Triggered Rule Discovery in Classifier Systems. In Schaffer (ed.) *Proceedings of the International Conference on Genetic Algorithms.* Morgan Kaufmann, pp 265–274.
8. Bull, L. (2004)(ed) *Applications of Learning Classifier Systems.* Springer.
9. Bull, L. & Kovacs, T. (2005)(eds) *Foundations of Learning Classifier Systems.* Springer.
10. Butz, M. & Wilson, S.W. (2002) An Algorithmic Description of XCS. *Soft Computing* 6(3-4): 144–153.
11. Crowley, M.F. & Field, R.J. (1986) Electrically coupled Belousov-Zhabotinskii oscillators 1. Experiments and simulations, *J Phys Chem.* 90: 1907–1915.
12. Crowley, M.F. & Epstein, I.R. (1989) Experimental and theoretical studies of a coupled chemical oscillator: phase death, multistability and in-phase and out of phase entrainment. J Phys Chem. 93: 2496–2502.
13. Dolnik, M. & Epstein, I.R. (1996) Coupled chaotic oscillators. *Physical Review E* 54: 3361–3368.
14. Dorigo, M. & Colombetti, M. (1998) *Robot Shaping.* MIT Press.
15. Field, R. J. & Noyes, R. M. (1974) Oscillations in chemical systems. IV. Limit cycle behavior in a model of a real chemical reaction. *J. Chem. Phys.* 60, 1877–1884.
16. Gorecki, J., Yoshikawa, K. & Igarashi, Y. (2003) On chemical reactors that can count. *J Phys Chem. A* 107: 1664–1669.
17. Hjelmfelt, A. & Ross, J. (1993) Mass-coupled chemical systems with computational properties. *J Phys Chem.* 97: 7988–7992.
18. Hjelmfelt, A., Schneider, F.W. & Ross, J. (1993) Pattern-recognition in coupled chemical kinetic systems. *Science* 260: 335–337.

19. Hjelmfelt, A., Weinberger, E.D. & Ross, J. (1991) Chemical implementation of neural networks and Turing machines. *PNAS* 88: 10983–10987.
20. Hjelmfelt, A., Weinberger, E.D. & Ross, J. (1992) Chemical implementation of finite-state machines. *PNAS* 89: 383–387.
21. Holland, J.H. (1975) *Adaptation in Natural and Artificial Systems.* University of Michigan Press.
22. Holland, J.H. (1986). Escaping brittleness: the possibilities of general-purpose learning algorithms applied to parallel rule-based systems. In Michalski, Carbonell, and Mitchell (eds) *Machine learning, an artificial intelligence approach.* Morgan Kaufmann.
23. Holz, R. & Schneider, F.W. (1993) Control of dynamic states with time-delay between 2 mutually flow-rate coupled reactors *J Phys Chem.* 97: 12239.
24. Jung, P., Cornell-Bell, A., Madden, K.S. & Moss, F.J. (1998) Noise-induced spiral waves in astrocyte syncytia show evidence of self-organized criticality. *Neurophysiol.* 79: 1098–1101.
25. Kádár, S., Amemiya, T. & Showalter, K. (1997) Reaction mechanism for light sensitivity of the Ru(bpy)(3)(2+) catalyzed Belousov-Zhabotinsky reaction. *Journal of Physical Chemistry A* 101 (44): 8200–8206.
26. Kawato, M. & Suzuki, R. (1980) Two coupled neural oscillators as a model of the circadian pacemaker. *J Theor Biol*, 86: 547–575.
27. Krug, H.-J., Pohlmann, L. & Kuhnert, L. (1990) Analysis of the modified complete Oregonator accounting for oxygen sensitivity and photosensitivity of Belousov-Zhabotinsky systems, *J. Phys. Chem.* 94, 4862–4866.
28. Kuhnert, L., Agladze, K.I. & Krinsky, V.I. (1989) Image processing using light sensitive chemical waves, *Nature* 337: 244–247.
29. Laplante, J.P., Pemberton, M., Hjelmfelt, A. & Ross, J. (1995) Experiments on pattern recognition by chemical kinetics. *J Physical chemistry* 99: 10063–10065.
30. Lebender, D. & Schneider, F.W. (1994) Logical gates using a nonlinear chemical reaction, *J. Phys. Chem.* 98: 7533–7537.
31. Michalewicz, Z. & Fogel, D. (1999) *How to Solve it.* Springer Verlag.
32. Motoike, I.N., Yoshikawa, K., Iguchi, Y. & Nakata, S. (2001) Real time memory on an excitable field. *Physical Review E* 63 (036220): 1–4.
33. Rossler, O.E. (1974) In M. Conrad, W. Guttinger & M. Dal Cin (eds) Lecture notes in Biomathematics. Springer, pp 399–418 and 546-582.
34. Sielewiesiuk, J. & Gorecki, J. (2002) Passive barrier as a transformer of chemical frequency. *Journal of physical chemistry A* 106: 4068–4076.
35. Steinbock, O., Kettunen, P. & Showalter, K. (1996) Chemical wave logic gates. *J Phys Chem*, 100: 18970–18975.
36. Steinbock, O., Toth, A. & Showalter, K. (1995) Navigating Complex Labyrinths: Optimal Paths from chemical waves. *Science* 267: 868–871.
37. Stuchl, I. & Marek, M. (1982) Dissipative structures in coupled cells: experiments. *J. Phys Chem.* 77: 2956–63.
38. Sutton, R. & Barto, A. (1998) *Reinforcement Learning.* MIT Press.
39. Toth, A., Gaspar, V. & Showlater, K. (1994) Signal transmission in Chemical Systems: Propagation of chemical Waves through capillary tubes. *J. Phys Chem.* 98: 522–531.

40. Wang, J., Kadar, S., Jung, P. & Showalter, K. (1999) Noise driven Avalanche behaviour in subexcitable media. *Physical Review Letters* 82: 855–858.
41. Watkins, C.J. (1989) Learning from Delayed Rewards. Ph.D. Thesis, Cambridge University.
42. Wilson, S.W. (1995) Classifier Fitness Based on Accuracy. *Evolutionary Computing* 3: 149–175.
43. Zaikin, A.N. & Zhabotinski, A.M. (1970) Concentration Wave Propagation in 2-Dimensional Liquid-Phase Self-Oscillating System. *Nature* 225 (5232): 535.

Existence and persistence of microtubule chemical trails — a step toward microtubule collision-based computing

Nicolas Glade

TIMC-IMAG CNRS UMR5525, Université Joseph Fourier – Grenoble, Faculté de Médecine, F-38700 La Tronche Cedex
Nicolas.Glade@imag.fr

Abstract. Microtubules and actin fibres are dynamic fibres constituting the cell skeleton. They show seemingly computational behaviours. Indeed, depending on conditions of reaction, they self-organise into temporal and/or spatial morphologies *in vitro* and act as micro-machine processors and actuators in living cells. I discuss here about their possible reuse to construct microtubule-based computational systems. I analyze particularly the possible existence of computational events based on 'chemical collision' between microtubules. A molecular model of microtubule disassembly has been developed to verify that heterogeneities of composition and/or concentration of the chemical medium can form from shrinking microtubules and persist. They could serve as an efficient communication channel between microtubules. Numerical simulations show that diffusing tubulin molecules can explore large distances (0.5 μm) during the disassembly of only one molecule. This leads to the formation of only very weak heterogeneities of composition. However, the model predicts that they could be more important due to microtubule arrays.

1 Introduction

Dynamic reacting fibres are numerous in nature. In living systems, they are, for instance, microtubules and actin fibres, both forming the living cells cytoskeleton, or helical tubular structures of some viral capsids such as those of T4 phages or tobacco mosaic viruses. Carbon nanotubes or

DNA-designed nanotubes [1] are examples in artificially produced systems.

Some of these systems, in particular microtubules, are able to self-organise from homogeneous 'soups of molecules' to well structured morphologies made of ordered populations of dynamic reacting fibres. Little is known about their self-organisation. It is thought to be out-of-equilibrium and to imply reaction and diffusion processes coming from the chemical or physical interactions of large collections of reacting fibres. Static interactions are also present and may reinforce self-organisation.

Such systems are studied for the consequences of their self-organisation on the function of living systems. Self-organisation of microtubules is proposed to be one way by which living cells are sensible to gravity or other external fields [2–4]. Actin fibres were also shown to self-organise. Liquid-crystalline processes were proposed to explain the phenomenon [5]. Associated with other proteins, they show different forms of self-organisation, in particular the formation of 'actin comets' propelling bacteria in infected cells [6, 7]. Moreover, an artificial system constituted of carbon nanotubes, in particular conditions of reaction, is also able to self-organise [8, 9]. They form coral like 'macroscopic' structures made of self-organised carbon nanotubes. These structures form dynamically and imply permanent processes of creation, growth (catalyzed by cobalt particles) and shrinkage (due to atomic hydrogen) of carbon nanotubes. The manner this system self-organises has similarities with that of microtubule solutions.

In all these systems, self-organisation results from the numerous individual interactions that occur between 'molecular agents'. They look to be computed. Information and its 'quasi algorithmic' processing are primordial principles of the biological function and organisation. Indeed, living matter acts since more than 3.5 billion years to develop the most efficient forms of information processing and nano-machines that could exist. It includes the exploitation of the informative DNA molecule, all organisation, transport and communication processes, and extends to most complex levels of information processing and nano-machinery as the function and organisation of the organisms, their reproduction, or the 'calculi' realised in the nervous system, bringing to the emergence of consciousness.

Such sources of inspiration abound in Nature. They motivate new research approaches and begin to be exploited in computational sciences. Their use has been thought since many years. For instance, cellular automata models were directly inspired from living systems. They were used in particular to answer the question of self-replication and were also shown to be able to realise computational tasks. However, molecular computation is relatively new and a consensus upon it was initiated by M. Conrad that furnishes its principal concepts [10]. Yet, he mentioned the importance of self-organisation in biological systems and cites Hameroff ideas about microtubules [11]. Because of their structure, their kinetic functioning — their ability to self-assemble — and their self-organisation properties, they are of great interest in computational science. Theoreticians proposed that computational events could occur inside microtubules or between them [11–18]. From this, it has been proposed that they could constitute a kind of 'cellular nano brain' able to intervene in particular in neuronal brain functions.

This article focuses on microtubule self-organisation and on the molecular processes that generate it. In particular I develop the aspect of the possible communication between microtubules by the way of the formation of tubulin trails. I emphasize that self-organisation could be viewed as the result of numerous computational events coupled to various communication channels. I also propose several manners to control this phenomenon. Notions such as programmability (structural control), adaptability, robustness of a computation with microtubule solutions, are discussed.

Microtubule solutions are particularly interesting because they are easy to produce, to use and to manipulate. They can be considered as a toy-model system usable to develop and test the concepts of computation with dynamic fibres. Moreover, they don't only exist in the form of a simulated program. They are real ! So they could be used to conceive real processing units. Nature has beaten us: cells and organisms are yet equipped with such chemical processors [17].

2 Microtubule self-organisation

Microtubules are tubular shaped supra molecular dynamic assemblies of about 25 μm diameter and several microns length. They are present in all eukaryotic cells, constituting, with actin fibres, the major part of their cytoskeleton. Cell shape, cell moves, intracellular transport and cell division are biological processes in which they play a substantive role. Moreover, they are important consumers of chemical energy (GTPases) and should probably participate greatly to energy regulation in living cells.

It's known since more than 15 years that *in vitro* solutions containing only purified microtubules and chemical energy, the GTP (guanosine triphosphate), can self-organise into macroscopic stationary morphologies [19, 20]. The forming patterns present periodic variations of concentration and orientation of microtubules [21]. They are also known to contain imbricated self-similar substructures on several scale levels [22]. These patterns form progressively, in several dozen minutes to hours, permanently consuming chemical energy to maintain the reaction out of the thermodynamic equilibrium [20]. The reaction implies only two molecules, the protein tubulin (the constituting brick of microtubules) and a nucleotide, the GTP (a chemical energy source allowing microtubules to disassemble). The reaction is initiated by warming the solution from 4 ° C to 35 ° C. Microtubules form within one or two minutes and, during hours, they reorganise progressively into these macroscopic well structured morphologies. Moreover, this phenomenon is sensible to weak external fields such as gravity [2,3,23], vibrations [26], magnetic [24–26] or electric fields [24, 27]. The intensity or the orientation of these fields can trigger and modify the manner that microtubules will organise.

A liquid-crystalline self-organisation mode like in actin birefringent solutions [5] can't be excluded [28–30]. Nevertheless, microtubules are highly dynamic assemblies. Reaction and diffusion processes should intervene in these processes. Tabony's team proposed that they could form from reaction-diffusion processes [3, 31]. One of the principal assumptions of the proposed model was that microtubules could communicate between each other via a chemical way. Indeed, microtubules, while growing or shrinking, are locally modifying the concentrations and the chemical nature of the free molecules present in the medium such as, at least, tubulin and energetic nucleotides. For example, while shrinking, a microtubule is liberating a chemical trail constituted by free tubulin molecules (they are inactive, associated with GDP (guanosine diphosphate), but they are rapidly regenerated into tubulin active molecules, associated with GTP). Another, more debatable, hypothesis has been made concerning the chemotactic ability of microtubules. They used this subtlety to explain why new microtubules preferentially 'choose' to grow in these chemical trails. Of course, microtubules are not equipped with sensors that guide them. Nevertheless, this can be understood in terms of probability of presence and of preferentially growing direction. The probability that another microtubule nucleates or develops into this recumbent region is higher than in other parts of the solution. Moreover, the probability that the new microtubule grow in the direction of the trail is higher than in other directions. Microtubules are highly reactive. If they form and grow in these trails, they are stable and continue to exist; if not (when they nucleate in the chemical trail and grow in another direction, going out of the trail) they could penetrate in an area of the solution with unfavourable conditions of reaction. Then, they would probably disassemble rapidly. Once formed, the tubulin trails extend in all directions by molecular diffusion and, with time, decrease in intensity until the solution is homogeneous. Then, microtubules just have to be in the neighbourhood of a trail to perceive its influence. Population of social insects, like ant colonies, are self-organising due to a communication between individuals by way of chemical trails called pheromone trails. The idea was that populations of chemically interacting microtubules could behave in same manner. Self-organised patterns of concentration and orientation of microtubules were obtained with this model. Populations of microtubules were self-organising by reaction-diffusion processes, at a microscopic level only [32] or at a macroscopic level also. The model was taking into account the triggering effect of the weak external fields [3, 14, 33]. Further, other teams made experimental observations showing evidences of such processes: new microtubules form and grow in the pathways let by other disassembling microtubules [19, 34, 35].

Other forms of microtubule self-organisation were observed. In other conditions of reaction, for example when the rate energy regeneration is low, one can observe periodic temporal oscillations of the bulk solution [36, 37], or periodic microtubule concentration waves travelling at high speed rates [19,38]. When adding microtubules associated molecules (e.g. molecular motors like dynein or kinesin) more complex dynamic structures can appear (e.g. vortices [39]).

To resume, microtubule self-organisation seems to be an underlying mechanism linked to microtubule collective dynamics. The process is highly sensitive to molecular factors acting directly on the dynamics of microtubular ends, such as ion concentrations (calcium, magnesium), or other molecules like taxol, colchicine, nocodazole, used particularly in tumour therapeutics, or acting on populations of microtubules such as microtubule associated proteins (MAPs). They are also sensitive to the presence or absence of external fields: gravity, vibrations, magnetic or electric fields, or geometric factors [40, 41], acting globally on the whole solution. Using light is also a good manner to act globally or locally (e.g. with the punctual UV beams sources of microscopes or with UV lasers) on these solutions by breaking microtubules into small microtubules, producing more reacting microtubular ends and changing their dynamics [42], or by acting locally via other molecules, for example by using caged GTP molecules [43]. All these factors could be used to guide microtubule self-organisation to an expected result.

3 Principles of a microtubule chemically-colliding computing system

Although no experimental prototypes of microtubule-based computers have been realised yet, number of experiments and numerical simulations have respectively shown or predicted macroscopic and microscopic behaviours compatible with a possible use in computational sciences. This section contains a review of theoretical ideas (and original ideas) on how microtubules could naturally compute. I also propose techniques to control them so as to obtain expected results. All these ideas about computation are only propositions and different manners to look at the microtubular system. However, each of them are reinforced by true data and experimental observations.

3.1 Microtubules as 'Turing machine'-type supramolecular assemblies

Turing machines are the formal representation of any computing process. They don't exist as physical objects but can be rewritten in any algorithmic programming language and implemented into von Neumann computers to realise calculations. The classic Turing machine is represented as a unique reading and writing head, moving and acting on a data tape (e.g. magnetic or optical data storage). The data tape contains a series of discreet finite states. The head also contains its own finite state. The program of a Turing machine can be expressed with 4 symbols: one for the initial state of the head, one for the output action (writing to the tape or moving on it), one for the data value in front of the head (reading), one for the final state of the head after processing. Depending on these values and related actions, the head reads and writes the tape, and moves on it until the machine can not move and change. During that, a process is realised by successive individual steps. The result is the new data of the tape.

In comparison (Fig. 1), microtubules are existing physical objects that can also be represented by an algorithmic programming language in a numerical model [10]. Their two reacting ends can be assimilated to two 'independent' Turing machines, they however can be linked internally by soliton waves [13, 16, 17, 44] and transport of ions inside the tube [45], and the body of the microtubule to an internal memory.

Microtubules form spontaneously from tubulin proteins associated with a nucleotide triphosphate, the GTP. When forming, the GTP is hydrolysed and a phosphate is liberated for each molecule of Tubulin-GTP assembled. Note that the hydrolysis of GTP is due to a structural change of tubulin after assembly. It is a consequence of the assembly. However, once formed, the tubulin-GDP is allowed to disassemble. GTP hydrolysis, during the microtubule assembly, is a necessary condition for the microtubule disassembly. Then, the main body of the microtubule is principally constituted of tubulin associated with GDP. Nevertheless, it's a little more complicated at the microtubular ends. The assembly or the disassembly of microtubules is not linear with the concentration of free tubulin because of the presence of 'microtubule caps'. They are small zones of variable length (the length of, at least, only one molecule of tubulin to several nanometres long) having chemical and structural differences with the body of the microtubule. Their precise nature — chemical (cap made of a buffering layer of tubulin-GTP or tubulin-GDP-P, with the phosphate not yet hydrolysed or liberated) and/or structural (open sheet shape for growing microtubules, inside-out curled filaments for shrinking microtubules) — is still debated [46–48] but all agree to say they are protective zones against microtubule disassembly at the growing ends, and, respectively, against assembly at the shrinking ends. They act as buffer zones, permitting a microtubule to survive to an unfavourable reactive event and can introduce delays in the changes of reactivity of the microtubular ends. Further, considering the computational approach, they introduce internal states at the tips of the microtubules. Then, as would do Turing machines, microtubular ends 'read' and 'write' in the chemical informative support, the medium, modifying locally the concentrations, and the compositions, of tubulin and nucleotides. The reacting states at their ends and the manner they will react are dependent to their own state, the chemical composition and the structure of the caps, and to the local concentration of their constituting bricks, tubulin-nucleotide molecular complexes, in the medium. If the end of a microtubule is in a stable state and if it is surrounded by a sufficient concentration of tubulin-GTP molecules, it can grow. In this case, it consumes the local molecules of tubulin-GTP, then creating a local depletion of the concentration of tubulin-GTP. On the contrary, if the end of the microtubule is in an unstable configuration and the chemical environment unfavourable, it disassembles, liberating locally tubulin-GDP molecules.

Moreover, when assembling from free tubulin-GTP, the microtubule stores in its body the information that was locally present in the medium, in the form of free molecules, and can restore it locally when disassembling. Indeed, when a microtubule disassembles, it leaves along its trajectory a concentrated trail of tubulin-GDP molecules. Before diffusion acts, homogenising the medium, information is locally restored. This information is available for other neighbouring microtubules. Note that the

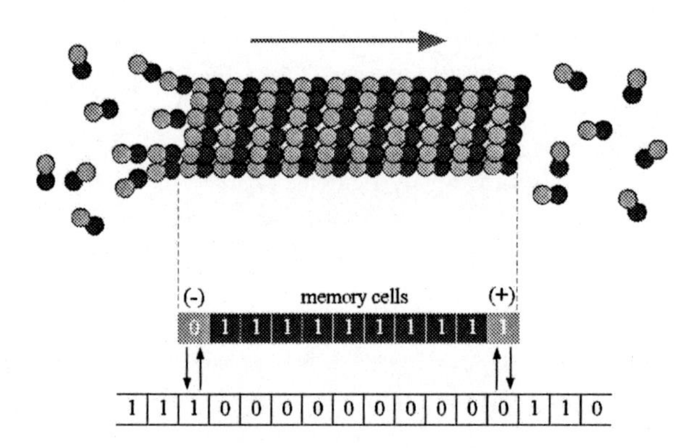

Fig. 1. Shows the scheme of a microtubule (up) growing at the right end and shrinking from the left end. At the ends, free tubulin heterodimers are diffusing. In parallel (down) is shown a microtubular Turing machine evolving on a 1D Boolean world (white cells). The concentrations of free tubulin heterodimers are idealised by Boolean values, 0 meaning 'no tubulin', 1 meaning 'presence of tubulin. The microtubule Turing machine is schematized by a non-reactive body, a series of 1 (black cells), acting as an internal memory of the previously encountered values in the world, and 2 writing and reading heads (gray cells). The processes at the heads depend on their internal state (0 for the shrinking end, 1 for the growing end) and on the local state encountered in the medium in front of the head. When moving (growing or shrinking), they modify locally the medium, writing a 1 (when liberating a 1) or a 0 (when consuming a 1).

storage of information into microtubule bodies has a cost: it consumes GTP to assemble. Information is completely restored when tubulin-GTP molecules are regenerated by exchange of the tubulin-associated GDP with free GTP nucleotides present in the medium.

3.2 Microtubular Turing machines compute self-organised morphologies

If microtubules are 'Turing machines'-type supramolecular assemblies, do they realise computational processes in microtubule solutions and how?

Computational processes occurring inside microtubules have been already proposed and theoretically studied. In particular, S. Hameroff and J. Tuszinsky initiated and have developed a framework for a possible quantum computation in microtubules and its eventual role in biological processes, in particular brain processes such as memory, signal processing and consciousness [13, 16–18, 44].

Here, I focus on more macroscopic phenomena, easiest to observe and to control, that could also realise computational tasks. I consider the manner by which collections of microtubule Turing machines are interconnected, producing local and global ordering. I propose that this could be understood as the result of numerous computational events. In time, solutions of microtubules reach asymptotic behaviours, temporal and/or spatial, maintained by a permanent consumption of chemical energy.

In general, in complex systems, when some individuals are close together, they can interact — directly or via their surrounding environment — and this affects their internal states and their behaviours. These simple events are elementary computation tasks, also called 'collision-based computing' [15]. They also could be associated to single instructions of a computer program. Correctly ordered and interconnected, these collision computing elements can realise more complex tasks. Ordering them to realise a particular task corresponds to a programming of the system.

From the concerted specific action of such computational elements can emerge a result in the form of a shape, a construction (social insects nests), a morphology (body patterning by pigmented cells [49]), or other emerging tasks (sorting out of objects and optimization tasks [50–53]). This may also be the case for microtubule populations. From their dynamics coupled to communications between them, a temporal [36, 37] or spatio-temporal [2, 19, 22, 31, 54] self-organising behaviour emerges.

Saying that self-organised asymptotic behaviours are 'results' of a computation brings up the question of the stopping problem. Indeed, in these systems, by definition, as long as they are self-organising, equilibrium is never reached, so the systems never stop on a particular result (the state of the system can always change). It's easier to look at them as real time adaptable processors. The morphologies that form correspond to the spatial stationary states of the chemical processor. They are maintained because of a permanent consumption of energy and reactants, and are waiting for external stimulations to change. In case of an n-states periodic temporal behaviour, the associated possible states of the chemical processor would be each of the n-states. When external stimuli are applied to (load in) the solution, its behaviour adapts itself to the new conditions producing new asymptotic states. When all energy or reactants is dissipated, all out-of equilibrium emergent behaviour disappears. The system then stops to compute but this is not equivalent to the stopping of a program.

Such behaviours can be controlled or biased to realise programmed tasks. Examples are known where populations of interconnected Turing machine type individuals, well ordered and sharing a common informational support, produce computational tasks. This is the case in cellular automata simulations such as the well known Conway's 'Game of Life', where individuals self-organise and compute, depending on their internal state and on local neighbouring states [15, 55, 56]. Experiments concerning the effect of external factors on microtubule spatially self-organising solutions have shown that the self-organising behaviour can be influenced to produce one type of morphology or another [19, 36, 37]. It's reasonable to think that the same could be done with only temporal or spatio-temporal microtubule self-organising solutions.

In both cases, self-organising system or pre-ordered (programmed) system, computation can occur because information is shared and exchanged between individual agents. In addition, any self-organising system that can be controlled is susceptible to be programmed.

At a collective level, microtubules are controllable. We 'only' have to understand how microtubules interact and how to control them finely. Let's look at the list of possible communication ways that could exist in microtubule solutions, allowing self-organisation to occur.

3.3 Inter-connexions between microtubules — communications in the solution

They are numerous manners by which microtubules can 'discuss' in a solution. These ways of communication are different in terms of rapidity, distance range, directionality and efficiency. Some are well established experimentally or are predicted by simulations; others are more hypothetic and hard to be observed.

- *Microtubules treadmilling* (and moves in general). *In vitro* or in living cells, microtubules are always-moving objects due to their dynamic nature. They have been observed to behave frequently in dynamic instability (stochastic-like behaviour of their reactive ends) or in treadmilling motion. During the former, they grow at one end and shrink at the other. A moving object is observed although the assembled tubulin does not move. While travelling, they behave as long range carriers of information in the solution, on the understanding that they can deliver this information locally after a long travel throughout the solution. This way of communication exists during all the life of the microtubule, is directional, but slow. It can't exceed the maximum growing rate of about 1 to 10 μm.min^{-1}.

- *Microtubular chemical trails or depletion areas: diffusion around the ends of the microtubule.* They will be more discussed in Section 4. By growing or shrinking, microtubules modify locally, around their reacting tips, the composition and/or the concentration of reactants. A possible formation of chemical trails behind microtubule shrinking ends and of depletion areas around growing ends have been stressed by several teams. They suggested and calculated that the activity of a microtubule could influence locally, or at a more important distance, the reactivity of neighbouring ones [3, 32, 56, 58]. Let in the pathway of microtubules, they should be directional. Their propagation by molecular diffusion is quick at the microtubule scale. If they exist, they would constitute one of the most fundamental mechanisms for the local microtubular inter-connexion. This can be called a chemical colliding computational event [15].

- *Molecular diffusion inside microtubules* [45]. Odde shown in 1998 that small molecules (ions, taxol, antibodies, nucleotides...) and tubulin heterodimers can diffuse inside the tube of microtubules. This was suspected because of the exclusive localization of interaction sites of molecules like taxol inside microtubules. Odde evaluated the diffusion for small molecules and for tubulin heterodimers. He showed that tubulin diffuses rapidly inside the microtubule, reaching the equilibrium inside of a 20 μm length microtubule in about 1 minute. What is important here is that this transport is purely directional, driven by the shape of the microtubule. When a molecule of tubulin explores a sphere of about 6 μm radius in 1 second outside microtubules, the same molecule would travel at the same rate but in only one direction inside the microtubule. This transport is directional, can be long range, depending on the size of the microtubule, and fast. Other molecules, such as ions or energetic nucleotides that modify the reactivity of microtubules, can also travel inside microtubules.

- *Avalanches of disassembly in solutions near to instability.* Molecular diffusion is fast at microtubule scale but is limited in range. Tubulin-GDP emitted by a microtubule and travelling far from the tip will be rapidly diluted in the population of GDP-tubulin molecules emitted by other microtubules. In return, locally, an inhibitive modification of the composition or the concentration of reactants (increase of the tubulin-GDP concentration or depletion of tubulin-GTP) would affect the nearest microtubules, causing their backhanded disassembly. If, at one moment of the reaction, numerous microtubules in the solution are in a close-to-disassembly state and their density number sufficient, the locally-emitted information of disassembly could propagate very quickly as in excitable media. Microtubules will function as amplifiers and relays of the signal. In self-organised solutions, microtubules are spaced by only one or two microtubule diameters [59]. It's a very short distance for the molecular diffusion. The propagation rate will only be limited by the reactive rates of each microtubule and their levels of instability. It will propagate very quickly at long distances if all the solution is in an instable state. The phenomenon will start from instable nodes (groups of microtubular ends) and will propagate all around. In spite of the global instability state, stability nodes, as to say small areas where microtubules groups are more stable than all around, can survive in the solution. The avalanche of disassembly encounters these regions of hindrance, is locally stopped by them, and progressively decreases in intensity. At a macroscopic level, this could produce very quickly microtubule concentration gradients over millimetres or centimetres. The macroscopic manifestation of this plausible process is observed in microtubule solutions *in vitro* where microtubule concentration waves can form and propagate at rates of about 1 mm.min^{-1} [19]. It implies that microtubule reactivity is synchronised by any process over long distances. This avalanche phenomenon can explain that as well as the formation, in only few minutes, of the 'longitudinal bar' of microtubules observed in spatially self-organised solutions [40]. At least two observations reinforce this suggested mechanism. First, moments of instability exist in microtubule self-organising solutions: in Tabony's solutions, a chemical instability occurs 6 minutes after the beginning of the reaction, a moment where microtubules are particularly concentrated and instable (the instability is followed by a phase of rapid disassembly) [60]. In Mandelkow's solutions, growing phases alternate locally with massive disassembling phases, producing microtubule concentration waves. Moreover, the fact that the concentration waves or the spatial self-organised stripes depart from, or are influenced by, the sample boundaries [3, 19, 38, 40, 41] is consistent with this suggested mechanism.

- *Quasi-particle (conformational solitons) waves inside microtubules* [16, 17, 44]. Although not yet proved experimentally, theoreticians suggested that tubulin heterodimers, as 'electrets' (electric dipoles), behave as bits of information, the microtubule lattice (the regular arrangement of tubulin in the body of the microtubule) becoming the place of internal computational events. Changes of the states of one or some heterodimers at one site of the microtubule propagate rapidly all around. Coherent states of neighbouring tubulin heterodimers can appear and propagate as soliton waves on the surface of the tube. Collisions of soliton waves can occur, giving rise to computational events directly at the surface of the microtubule. Ingoing and outgoing information comes from exchanges with the outer medium and with the neighbouring microtubules by way of local external fields (electromagnetic fields). Indeed, the formation of zones of coherent states on microtubules generates local oriented fields (perpendicular to the surface of the microtubules) that can be perceived by other microtubules. Moreover, it has been predicted that incoming soliton waves could affect the reactivity of microtubular ends. Reciprocally they could be generated by changes at microtubular ends (converting the energy of GTP into soliton waves). In this model, microtubule associated proteins would have an important role maintaining microtubules close together. Such a mechanism would allow the information to travel very rapidly on an individual microtubule from one end to the other (linking them by a propagation of internal states in the microtubule). Individual microtubules would then behave as nerves, receiving and integrating a signal at one end and/or along the body, propagating it, and emitting changes at the other end by the way of proportional changes of reactivity.

- *Mechanical interactions mediated by MAPs.* Microtubule associated proteins play an important role in living cells, controlling the precise structure of the microtubule cytoskeleton. They are of different types, producing various effects. Most of the time, they are molecular motors using energy of triphosphated nucleotides to travel along microtubules to one direction or to the other. Recent works shown that adding MAPs to microtubule solutions produces well ordered patterns [39]. As additive instructions to a programming language, they increase the diversity of behaviours of the microtubular system.

3.4 Control of the process

The self-organisation phenomenon is autonomous and adaptable to external stimuli. It could then be used directly as an enslaved system, controlled by an interaction loop with the environment. Nevertheless, in regard to the huge combinations of elementary computational events in microtubule solutions, it would be interesting to consider their programmability.

The von Neumann computer opts for structural programmability [10]. That means that two programs having exactly the same structure, initiated with the same set of data, will produce the same result. On the contrary, in a molecular computer, structural programmability is difficult to implement. Programming the system consists in the initiation of its configuration by setting exactly all states, positions and orientations of the molecules in the system. For the microtubules, that means to organise them precisely. Giving that, excepted in numerical simulations, the exact programming of the real molecular computer will not ensure to obtain the same result each time. Genuinely, due to thermal agitation and to the stochastic nature of chemical reactions, two identically-programmed processes will rapidly diverge. Two microtubules, in the same conditions, will not assemble or disassemble in the same way. So, controlling them is probably not interesting because the result of their interaction is weakly deterministic.

However, when considering a population, even small, of individuals, it may not be completely non-deterministic. If it was the case, we would not obtain reproducible self-organised behaviours. Two self-organising processes, initiated with similar conditions of reaction, will not give identical but very similar behaviours (morphologies, waves ...).

Solutions exist to consider structural programmability of microtubule solutions. They will be more discussed in the concluding remarks. In brief, they could consist in the use of the advantages of self-organisation, in its control, and in the isolation of small quantities of self-organising populations of microtubules. Micro-volumes of solution can be isolated physically, using physical boundaries, or virtually, using local actions (e.g. local electric fields delivered by electrode arrays). In fact, they self-organise in very small samples of cellular dimensions, as PDMS designed micro test tubes, micro-capillaries, phospholipidic vesicles and solution droplets [41,61].

For instance, a structural programming of the system would be to use these micro-samples as micro-chips, controlling locally their individual self-organisation and configuring the connections of a population of these chips as could be done in cellular automata. Each chip will behave independently, reaching progressively its own morphological or temporal attractor, and will be influenced by information received from the neighbouring connected chips. Self-organisation of microtubules, in each chip, is controllable by external fields such as gravity or vibrations, magnetic, electric and probably electromagnetic fields, by temperature variations applied on each micro-sample, by UV-light ... Controlling the shape and the dimensions of each sample, and the channels connecting them, is also a good manner to control the computation. In living tissues, similar situations occur. For example, in the cardiac tissue, excitation waves are initiated tissue by changing the shape of one of the cells by local application of a pressure [62].

4 Simulation of the formation of tubulin-GDP 'trails': chemical communication between microtubules is possible

The chemical activity of microtubules causes the formation of local variations of concentration and composition of the chemical medium around their reacting tips. This assumption is intuitive. It has been proposed repeatedly that the formation of such variations could influence microtubule dynamics and self-organisation. In 1990, Robert et al published a simple chemotactic model of microtubule self-organisation where individual microtubules coordinate each other and self-organise, following the gradients of tubulin concentration, themselves modified by the activity of the microtubules [32]. The general idea was there but the influence of the activities of the microtubular ends on other microtubules, by the intermediate of free tubulin concentration variations, remained hypothetic.

A quantitative model of formation of a depleted area of free GTP-tubulin around the growing tip of a microtubule was proposed by Odde [58]. The motivation of this study was to estimate if the formation of this zone and its homogenisation by diffusion could be a limitation for the growth reaction. The result, an analytical solution of a reaction-diffusion equation, estimated that for a microtubule growing at 7 μm.min^{-1}, the concentration at the tip was 89% of the concentration far from the tip and that the concentration gradient was extending to less than 50 μm from the tip (about 1-2 microtubule diameters). The model proposed by Glade et al [3] also predicted the formation of similar depleted areas at the growing ends of microtubules with the formation of tubulin concentrated trails at the shrinking ends (Fig. 2). This local phenomenon was hypothesised to be the most fundamental manner by which microtubules communicate. It is reinforced by observations *in vitro* [19] and *in vivo* [34] showing that new microtubules preferentially grow in the trajectory of shrinking ones.

Nevertheless, these models were simulating continuous quantities of tubulin molecules, expressed in μM, that way approximating the reality. In real solutions only few molecules diffuse from the reacting ends of microtubules and are diluted in large amounts of free tubulin. In consequence, the differences of tubulin amounts surrounding a shrinking end are very small. In these conditions, it could call into question the molecular explanation, the communication between microtubules mediated by free tubulin molecules, of microtubule self-organisation [3, 32]. I wanted to verify the possible existence of such tubulin 'trails' and their survival inversely depending on diffusion. I also wanted to answer the question of whether spatial variations of tubulin amounts were detectable by other microtubules and if it was the case, how?

I conceived a molecular model where all tubulin molecules were represented. Tubulin heterodimers are approximated by ellipsoids of 8 x 4 x 4 nm diameters. Their translational and rotational diffusion rates along the 3 axis where calculated from their shape, the 3 radii of the ellipsoid, or from the macroscopic diffusion constant of tubulin in the cytoplasm [45, 58, 63]. Using the Stoke's relationship, corrected for a prolate ellipsoid (the approximated shape of tubulin heterodimers), I obtained the following. In water, for the X (great axis) and Y (short axis Y=Z) axis of the heterodimers, the translational rates are respectively 5.7 nm.μs^{-1} and 4.0 nm.μs^{-1}, and the rotational rates 11.1°.ns^{-1} and 2.0°.ns^{-1}. In

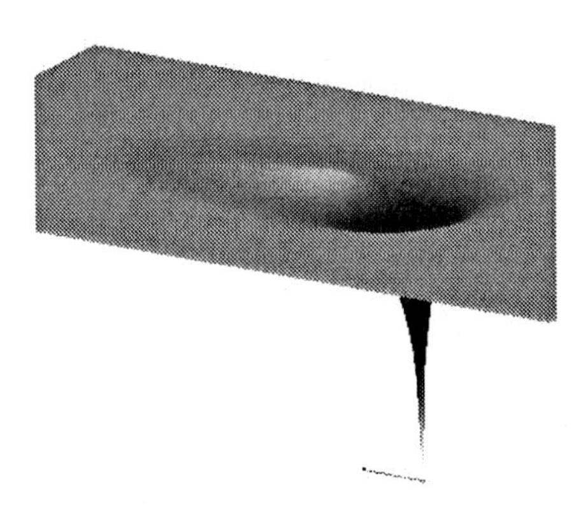

Fig. 2. A tubulin trail as simulated by the spatiotemporal differential equation system of a reaction-diffusion 2d-model similar to that proposed by Glade et al [3]. The figure shows the tubulin-GTP concentration map (represented in false 3D and grey scales), in superposition of a microtubule (above). In this simulation, an isolated microtubule is in treadmilling motion, growing at the right end and shrinking at the left one. At the growing end, an 'intense' depletion of tubulin forms (the black hole). At the shrinking end, tubulin-GDP is liberated and rapidly converted into tubulin-GTP producing a concentrated trail (white trail at the left of the depletion). The microtubule measures 10 μm.

the cytoplasm, the translational rate values are respectively 2.0 nm.μs^{-1} and 1.4 nm.μs^{-1}, and 1.4 $^\circ$.ns^{-1} and 0.24 $^\circ$.ns^{-1} for the rotational rates. The time step was fixed at 2.5 ns for the most precise simulations (to quantify the macroscopic diffusion) and at 5 μs for the others. At each time step, all molecules diffused randomly according to these values, with the constraint that 2 molecules couldn't exist in the same place. From these simulations, I verified the measured macroscopic diffusion of the population of tubulin heterodimers. It was perfectly consistent with the value measured experimentally (5.9 10^{-12} $m^2.s^{-1}$ in the cytoplasm and 4.9 10^{-11} $m^2.s^{-1}$ in water). The boundary conditions were toric for tubulin-GTP molecules, to maintain their density in the sample, and permeable for tubulin-GDP molecules, to obtain their correct density profiles.

The objective was to observe the diffusion of the liberated molecules of GDP-tubulin from the tip of a shrinking microtubule. Preformed microtubules, with a space step between tubulin heterodimers of 0.57 nm along the X axis, and an angle step of 27.69 $^\circ$ between two successive heterodimers, were designed.

To test the formation of a tubulin trail, I placed an isolated microtubule of 0.5 μm long at the centre of a simulated cubic sample of 1.4 μm side, oriented along the X axis, in a medium containing 1 μM tubulin-GTP (5400 molecules.μm^{-3}).[1] The microtubule was allowed to disassemble constantly. I simplified the microtubule disassembly considering the liberation of GDP-tubulin molecule by molecule and not the liberation of tubulin coiled oligomers from proto-filaments. Assembly was not permitted. The conversion of tubulin-GDP into tubulin-GTP was not either permitted.

In this simulation, the microtubule was disassembling 100 x faster than a real microtubule (normally shrinking at a maximum of about 20 $\mu m.min^{-1}$, as to say at a rate equivalent to 2000 $\mu m.min^{-1}$. In this case, an eye-observable concentrated area formed around the shrinking end (Fig. 3). At realistic rates of disassembly (20 $\mu m.min^{-1}$) however, the variation of molecular density around the tip was hard to detect, in particular in water. Indeed, molecular diffusion is a very fast process compared to microtubule disassembly. During the disassembly of one molecule in the cytoplasm, the previously liberated molecule has a sufficient time (about 1.8 ms) to explore an average sphere of 150 nm of radius (0.5 μm in water). In consequence, during microtubule disassem-

[1] Note that tubulin concentrations of 100 μM or the presence of numerous microtubules distributed in the sample don't really affect the diffusion rate of free tubulin (results not shown).

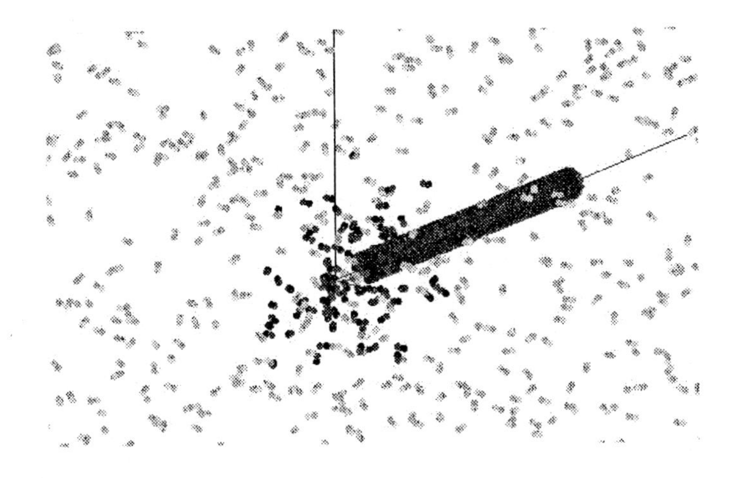

Fig. 3. Concentrated area of tubulin-GDP (black) in formation at the shrinking end of a microtubule. Tubulin-GTP heterodimers are displayed in grey. In this simulation, the microtubule disassembles 100 x faster (at 2000 μm.min^{-1}) than a normal microtubule.

bly - or assembly [58] - the solution is rapidly homogenous at the micron scale (largely greater than microtubules scales). Nevertheless, although it's not intense, the very weak gradient of GDP-tubulin molecules can be measured (Fig. 4). Globally, the GDP-tubulin heterodimers are nearest to the tip than far away.

This can be better observed by allowing several group microtubules to disassemble. I simulated 5 parallel aligned microtubules disassembling at 20 μm.min^{-1}. This time, the formation of a gradient of GDP-tubulin was clearly eye-observable and measurable. After 11 ms of reaction in cytoplasmic conditions, the gradient extends to about 0.7 μm from the tips of the microtubules and has an average maximum value of 4.5$\pm$1.8 molecules at the tips of the microtubules (Fig. 4, left).

In water, the intensity of the gradient is weaker. After 11 ms, the gradient of tubulin-GDP extends more rapidly to a distance of about 1.5 μm from the tips of the microtubules. It has an average maximum value of 1.31$\pm$1.08 molecules at the tips of the microtubules (Fig. 4, right).

In conclusion, microtubules, by their assembling or disassembling activity can produce local chemical heterogeneities. Unfortunately, even if dozens of tubulin-GDP-producing microtubules are located in the same place, the increase of total tubulin concentration is undetectable: the variation of density of tubulin-GDP is largely lower than the natural

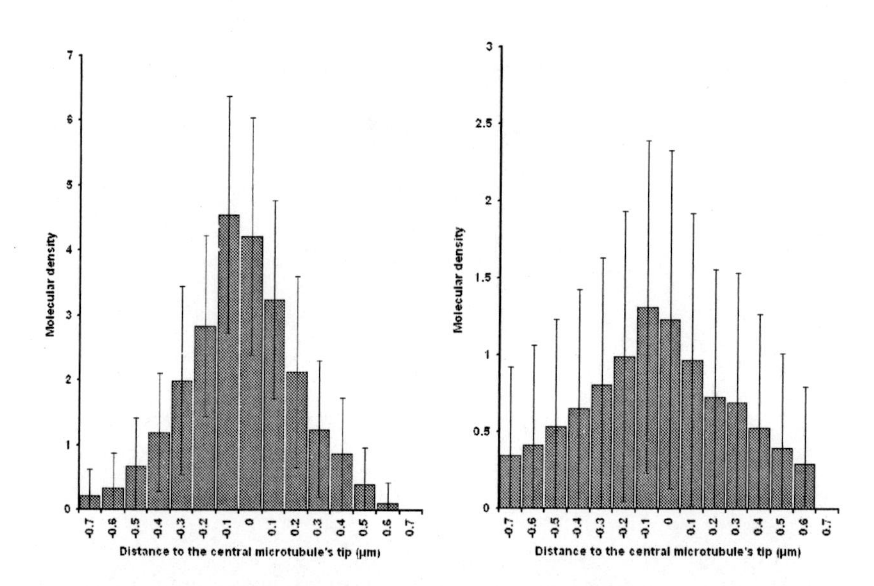

Fig. 4. (**Left**) Profile tubulin-GDP density around the tips of disassembling microtubules, measured from the centre of an array of 5 disassembling microtubules, each of them respectively separated by 30 nm (one microtubular diameter). All microtubules disassemble simultaneously at 20 μm.min^{-1} (0.54 heterodimer.ms^{-1}). The translational diffusion rate of individual molecules has been determined from the measured value of tubulin diffusion in the cytoplasm (5.9 10^{-12} m^2.s^{-1}) [63]. The quantity of tubulin-GDP molecules liberated is very low. In consequence integration in time is necessary to obtain its density profile. It has been reconstructed by integration of the density maps of 6 independent simulations, during 1.8 ms, between the simulation times 9.2 ms and 11 ms, along the 3 axis (a total of 6642 profiles). (**Right**) The same with a different diffusion rate. Here, translational diffusion rate of individual molecules has been determined from the shape of tubulin, using the viscosity of water at 37°C. The average value of the global diffusion constant obtained (4.9 10^{-11} m^2.s^{-1}) is 8x larger than in the cytoplasm, as measured in water [64]. The resulting density profile of tubulin is weaker but detectable.

fluctuations of total tubulin density (there are about 1000 molecules of tubulin-GTP per molecule of liberated tubulin-GDP). These heterogeneities are not concentrated areas, but composition modified areas.

For individual microtubules, they are very weak and extended in space, so, particularly in water-based solutions, it is improbable that an individual microtubule can influence another one. The effect is probably sensible when produced by a group of synchronously reacting microtubules. To produce an intense composition or concentration heterogeneity, there are two possible scenarios: (1) in a solution of randomly distributed microtubules, nodes of microtubular ends naturally exist that can form initial composition (or concentration) heterogeneity nodes; (2) microtubule arrays form locally by another mechanism (for example by electrostatic interactions). Then, the formation of heterogeneity nodes can provoke the recruitment or inhibition of microtubules as proposed in [3, 32]. Moreover, because of the rapid diffusion of tubulin molecules in comparison to the reactivity of the microtubules, tubulin 'trails' are not directional as proposed in the reaction-diffusion model of Glade et al [3].

Other effects could reinforce a little the intensity of tubulin-GDP concentrated areas. Indeed, microtubules can liberate individual heterodimers of tubulin-GDP but also oligomers of several assembled tubulin-GDP heterodimers. Such oligomers are N times longer than N separated tubulin-GDP heterodimers. Their diffusion rate is then approximately N times lower than that of free tubulin. This could help to maintain the free tubulin-GDP more concentrated in the neighbourhood of the tip. Further, during disassembly, short assembly events can occur sometimes and convert locally tubulin-GTP into tubulin-GDP immediately liberated and added to the concentrated pool at the tip.

5 Concluding remarks

Is it utopia to think microtubule solutions in terms of computing systems?

Writing this article was something like revisiting the works concerning microtubule dynamics and self-organisation. Nevertheless, the reader will have noted down that I don't give any plans to construct a microtubule-based computer. Indeed, reconsidering microtubule solutions and placing them in the context of molecular computation was similar to take to pieces an electronic computer and, looking at each piece, wondering what is its function and how it works.

Several elements of a possible computation are present in microtubule self-organising solution, and more generally in dynamic fibre populations. Now it's an inverse problem: with these pieces and tools, we have to reconstruct, and first, figure out the computer. Theoretical basis exist that describe what could be a reaction-diffusion computer [65] or a chemical collision based computer with reacting polymers [15]. However, a real implementation is not trivial.

A chosen approach could be to consider only the macroscopic level of self-organisation — generated morphologies or global behaviours — as done in other works to control external systems by an environment interaction loop [66, 67]. As in [67] it could be easy to use Mandelkow's microtubule solutions, with travelling waves of microtubule concentration, instead of a Belousov-Zhabotinsky (BZ) reaction, with a similar environment interaction loop, to control the behaviour of any *Animats*. It would probably not provide any additional advantage in comparison to BZ processors. Indeed, they react faster and are easier to realise.

I think there's something more powerful that can be extracted from microtubule systems and used, due to their particular dynamics and to the numerous communication ways that coexist in these solutions. The best would be to be able to program these systems at the molecular level, as to say, to initiate the microtubule processing system by placing microtubules (and other molecules) in such an arrangement (and state) that they compute in accordance with an known algorithm. The problems of structurally programmed molecular computers were underlined by Conrad [10]. Imagining that it's possible to set the initial states and connections at the beginning of the reaction, due not only to non-deterministic molecular agitation but also to an uncertainty of the control of this initialization, the local computing events would rapidly diverge and produce a result far from the expected one at the macroscopic level.

The two major problems are the precision of the final result and the manner to control the system with the maximum of accuracy and at the smallest scale.

Microtubules are not as sophisticated as neurons. They don't self-organise as precisely as neurons do to form a functional tissue realising a computation. The structural programming of a microtubule brain will then be delicate. Nevertheless microtubule solutions could learn more or less as would do neural networks, if the reactive conditions allow them to self-organise and to reorganise after an external stimulation (the asymptotic behaviour must not be too much stable). Tabony's conditions of self-organisation are too much stable: the produced stationary morphologies are difficult to modify after the 6th minute of the process that corresponds to the chemical instability, when the solution is sensible to external factors. Moreover, this process is very slow. Something closer to Mandelkow's conditions (more reactive) would be more appropriated. Once controlled, self-organisation would be directly used to initiate a 'more or less well programmed' processor that would realise quite well an expected calculus. The asymptotic behaviour would be refined by a learning feedback loop adjusting self-organisation, and so on.

Moreover, the feasibility of a fine control of a reaction-diffusion system to create computing elements has been shown on other systems, in particular with single chemical waves of the BZ reaction [65, 68]. A proposition would be to create small elements of computation, well localised in containers like vesicles, droplets or PDMS micro-tubs, containing the minimal amount of microtubules. Thereafter, they could be inter-connected like small electronic chips. The computation would emerge from the coupled dynamics of the individual chips.

Most of the methods presented before to control microtubule self-organisation (magnetic fields, vibrations, gravity and temperature ...) act on the entire sample without any discrimination in space. A fine control is possible using functionalized elements added to the system. A combination of different microtubules associated proteins is the most simple. Depending of the composition of the MAPs mix (type and concentration of the MAPs used), one can obtain a variety of self-organised morphologies [39]. MAPs can be added as free molecules in the solution, or be part of a functionalized surface (for example a surface patterned with MAPs). More powerful, microtubules functionalized with magnetic nanoparticles [69] allow controlling the positioning of individual microtubules with local weak magnetic field. Small populations of microtubules could be localised in vesicles or micro-samples elaborated in silicon polymers [41] and interconnected by other channels of communication like the gap junctions in the cells [16]. Finally, if microtubules are not the best fibres to do computation (their lifetime is short, about 1 or 2 days, and they react slowly), one could imagine to use other dynamic nano-fibres such as dynamic behaving carbon nanotubes [8, 9] or specially designed fibres [1].

In this article, I presented the beginning of a study concerning the existence of elementary computational events in microtubule solutions, in particular a communication between microtubules via 'chemical composition heterogeneities'. The logical continuation of this study is to verify by similar numerical simulations whether or not microtubular ends are sufficiently sensible to these variations of composition in the medium and that this mechanism can drive a certain self-organisation. The next step is to simulate and quantify the variation of the chemical states of two ore more microtubules during the chemical collisions that certainly occur in microtubule solutions *in vitro* or in living cells. This will depend on angle and distance parametres between microtubules. Moreover, due to thermal agitation, microtubular ends are always moving. This introduces a fuzzy term in the chemical interaction between microtubules.

Seeing natural processes as computing processes gives to biologists and physicists a different point of view and sometimes helps to break accepted paradigms. Microtubules serve as simple mechanical elements in the cell structure (tensegrity models of the cytoskeleton). In addition, the cytoskeleton is acting as an autonomous system sensible to external stimuli, conferring very complex behaviours to the cell. Comparing behavioural phenomena at the cellular scale and those that exist at the level of entire organisms is revealing. The motion of a migratory cell looks similar to that of an octopus, an organism more complicated. Flagella organelles are primitive caudal fins of swimming cells (e.g. spermatozoids). The bacteria *Listeria* reuse the actin cytoskeleton as a propelling motor in infected cells [6, 7]. These simple organisms developed many techniques to reuse the cytoskeleton, creating that way micro-machines. In more simple, non living, systems, microtubule activity and/or its self-organisation is intrinsically capable to act as a micro-machine, directly in a solution [54] or in phospholipidic vesicles [41, 61].

Over the aspect of micro-machinery, it's a temptation to imagine the role of primitive brain it could have had in simple organisms (e.g. ciliate organisms) during millions years of evolution. To end the article, I can't refrain citing these words of C. S. Sherrington, yet mentioned by Hameroff and Tuszynski in [17], about the 'seemingly intelligent' behaviours of single cell protozoa: "*of nerve there is no trace, but the cytoskeleton might serve*".

Acknowledgements

I am very much in debt to J. Demongeot (TIMC-IMAG, CNRS) for supporting the project, for all stimulating discussions and his helpful suggestions. I also want to thank A.-M. Bonnot (LEPES, CNRS) for the discussions about his self-organising system of dynamic carbon nanotubes and applications.

References

1. Rothemund, P. W., Ekani-Nkodo, A., Papadakis, N., Kumar, A., Fygenson, D. K., Winfree, E., Design and Characterization of Programmable DNA Nanotubes. J. Am. Chem. Soc. **126** (2004) 16345-16352. +supplement (S1-S19)
2. Tabony, J., Job, D., Gravitational Symmetry Breaking in Microtubular Dissipative Structures. Proc. Natl. Acad. Sci. USA **89** (1992) 6948-6952
3. Glade, N., Demongeot, J., Tabony, J., Numerical Simulations of Microtubule Self-Organisation by Reaction-Diffusion. Acta Biotheor. **50** (2002) 239-268
4. Crawford-Young, S. J., Effect of Microgravity on Cell Cytoskeleton and Embryogenesis. Int. J. Dev. Biol. **50** (2006) 183-191
5. Coppin, C. M., Leavis, P. C., Quantitation of Liquid-Crystalline Ordering in F-Actin Solutions. Biophys. J. **63** (1992) 794-807
6. Alberts, J. B., Odell, G. M., In Silico Reconstitution of Listeria Propulsion Exhibits Nano-Saltation. PLOS Biol. **2** (2004) e412
7. Boukella, H., Campas, O., Joanny, J. F., Prost, J., Sykes, C., Soft Listeria: Actin-Based Propulsion of Liquid Drops. Phys. Rev. E Stat. Nonlin. Soft Matter Phys. **69** (2004) e061906
8. Bonnot, A. M., Deldem, M., Beaugnon, E., Fournier, T., Schouler, M. C., Mermoux, M., Carbon Nanostructures and Diamond Growth by HFCVD: Role of the Substrate Preparation and Synthesis Conditions. Diam. Rel. Mat. **8** (1999) 631-635
9. Bonnot, A. M., Smria, M. -N., Boronat, J. F., Fournier, T., Pontonnier, L., Investigation of the Growth Mechanisms and Electron Emission Properties of Carbon Nanostructures Prepared by Hot-Filament Chemical Vapour Deposition. Diam. Rel. Mat. **9** (2000) 852-855
10. Conrad, M., On Design Principles for a Molecular Computer. Comm. ACM. **28** (1985) 464-480
11. Hameroff, S. R., Watt, R. C., Information Processing in microtubules. J. Theor. Biol. **98** (1982) 549-561
12. Tuszynski, J. A., Trpisov, B., Sept, D., Sataric, M. V., The Enigma of Microtubules and their Self-Organizing Behavior in the Cytoskeleton. Biosystems **42** (1997) 153-175
13. Tuszynski, J. A., Brown, J. A., Hawrylak, P., Dieletric Polarization, Electrical Conduction, Information Processing and Quantum Computation in

Microtubules. Are they Plausible? Phil. Trans. R. Soc. Lond. A **356** (1998) 1897-1926

14. Tuszynski, J., Sataric, M. V., Portet, S., Dixon, J. M., Physical Interpretation of Microtubule Self-Organization in Gravitational Fields. Phys. Lett. A **340** (2005) 175-180
15. Adamatzky, A., Collision-Based Computing in Biopolymers and Their Cellular Automata Models. Int. J. Modern Physics C. **11** (2000) 1321-1346
16. Hameroff, S., Nip, A., Porter, M., Tuszynski, J., Conduction Pathways in Microtubules, Biological Quantum Computation, and Consciousness. Biosystems **64** (2002) 149-168
17. Hameroff, S. R., Tuszynski, J. A., Search for Quantum and Classical Modes of Information Processing in Microtubules: Implications for the Living State. In: Musumeci, F., Ho, M. W. (eds.): Bioenergetic Organization in Living Systems. Proceedings of the Conference: Energy and Information Transfer in Biological Systems, Acireale, Italy. World Scientific, Singapore, (2003) 31-62
18. Faber, J., Portugal, R., Rosa, L. P., Information Processing in Brain Microtubules. Biosystems. **83** (2006) 1-9
19. Mandelkow, E., Mandelkow, E. M., Hotani, H., Hess, B., Muller, S. C., Spatial Patterns from Oscillating Microtubules. Science **246** (1989) 1291-1293
20. Tabony, J., Job, D., Spatial Structures in Microtubular Solutions Requiring a Sustained Energy Source. Nature **346** (1990) 458-451
21. Tabony, J., Job, D., Microtubular Dissipative Structures in Biological Auto-Organization and Pattern Formation. Nanobiology **1** (1992) 131-147
22. Tabony, J., Vuillard, L., Papaseit, C., Biological Self-Organisation and Pattern Formation by Way of Microtubule Reaction-Diffusion Processes. Adv. Complex Systems **3** (2000) 221-276
23. Papaseit, C., Pochon, N., Tabony, J., Microtubule Self-Organization is Gravity Dependant. Proc. Natl. Acad. Sci. USA **97** (2000) 8364-8368
24. Vassilev, P. M., Dronzine, R. T., Vassileva, M. P., Georgiev, G. A., Parallel Arrays of Microtubules Formed in Electric and Magnetic Fields. Bioscience Rep. **2** (1982) 1025-1029
25. Glade, N., Tabony, J., Brief Exposure to Magnetic Fields Determine Microtubule Self-Organisation by Reaction-Diffusion Processes. Biophys. Chem. **115** (2005) 29-35
26. Glade, N., Beaugnon, B., Tabony, J., Ground Based Methods of Attenuating Gravity Effects on Microtubule Preparations Show a Behaviour Similar to Space Flight Experiments and that Weak Vibrations Trigger Self-Organisation. Biophys. Chem. **121** (2005) 1-6
27. Stracke, R., Bohm, K. J., Wollweber, L., Tuszynski, J. A., Unger, E., Analysis of the Migration Behaviour of Single Microtubules in Electric Fields. Biochem. Biophys. Res. Commun. **293** (2002) 602-609
28. Hitt, A. L., Cross, A. R., Williams Jr., R. C., Microtubule Solutions Display Nematic Liquid Crystalline Structures. J. Biol. Chem. **265** (1990) 1639-1647

29. Baulin, V. A., Self-Assembled Aggregates in the Gravitational Field: Growth and Nematic Order. J. Chem. Phys. **119** (2003) 2874-2885
30. Ziebert, F., Zimmermann, W., Pattern Formation Driven by Nematic Ordering of Assembling Biopolymers. Phys. Rev. E. **70** (2004) 022902 1-4
31. Papaseit, C., Vuillard, L., Tabony, J., Reaction-Diffusion Microtubule Concentration Patterns Occur During Biological Morphogenesis. Biophys. Chem. **79** (1999) 33-39
32. Robert, C., Bouchiba, M., Robert, R., Margolis, R. L., Job, D., Self-Organization of the Microtubule Network. A Diffusion Based Model. Biol. Cell. **68** (1990) 177-181
33. Portet, S., Tuszynski, J. A., Dixon, J. M., Sataric, M. V., Models of Spatial and Orientational Self-Organization of Microtubules under the Influence of Gravitational Fields. Phys. Rev. E **68** (2003) epub 021903
34. Keating, T. J., Borisy, G. G., Centrosomal and Non-Centrosomal Microtubules. Biol. Cell. **91** (1999) 321-329
35. Shaw, S. L., Kamyar, R., Ehrhardt, D. W., Sustained Microtubule Treadmilling in Arabidopsis Cortical Arrays. Science **300** (2003) 1715-1718
36. Pirollet, F. Job, D., Margolis, R. L., Garel, J. R., An oscillatory Mode for Microtubule Assembly. EMBO J. **6** (1987) 3247-3252
37. Carlier, M-F., Melki, R., Pantaloni, D., Hill, T. L., Chen, Y., Synchronous Oscillations in Microtubule Polymerisation, Proc. Natl. Acad. Sci. USA **84** (1987) 5257-5261
38. Sept, D., Model for Spatial Microtubule Oscillations. Phys. Rev. E **60** (1999) 838-841
39. Surrey, T., Nedelec, F., Leibler, S., Karsenti, E., Physical Properties Determining Self-Organization of Motors and Microtubules. Science **292** (2001) 1167-1171
40. Tabony, J., Glade, N., Papaseit, C., Demongeot, J., Microtubule Self-Organisation and its Gravity Dependence. Adv. Space Biol. Med. **8** (2002) 19-58.
41. Cortes, S., Glade, N., Chartier, I., Tabony, J., Microtubule Self-Organisation by Reaction-Diffusion Processes in Miniature Cell-Sized Containers and Phospholipid Vesicles. Biophys. Chem. **120** (2005) 168-177
42. Walker, R. A., Inou, S., Salmon, E. D., Asymmetric Behaviour of Severed Microtubule Ends After Ultraviolet-Microbeam Irradiation of Individual Microtubules in Vitro. J. Cell Biol. **108** (1989) 931-937
43. Marx, A., Jagla, A., Mandelkow, E., Microtubule Assembly and Oscillations Induced by Flash Photolysis of Caged-GTP. Eur. Biophys. J. **19** (1990) 1-9
44. Georgiev, D. D., Papaioanou, S. N., Glazebrook, J. F., Neuronic System Inside Neurons: Molecular Biology and Biophysics of Neuronal Microtubules. Biomed. Rev. **15** (2004) 67-75
45. Odde, D.: Diffusion Inside Microtubules. Eur. Biophys. J. **27** (1998) 514-520
46. Mandelkow, E. -M., Mandelkow, E., Milligan, R. A., Microtubule Dynamics and Microtubule Caps: A Time Resolved Cryo-Electron Microscopy Study. J. Cell Biol. **114** (1991) 977-991

47. Caplow, M., Fee, L., Concerning the Chemical Nature of Tubulin Subunits that Cap and Stabilize Microtubules. Biochemistry **42** (2003) 2122-2126
48. VanBuren, V., Cassimeris, L., Odde, D. J., Mechanochemical Model of Microtubule Structure and Self-Assembled Kinetics. Biophys. J. **89** (2005) 2911-2926
49. Suzuki, M., Hirata, N., Kondo, S., Travelling Stripes on the Skin of a Mutant Mouse. Proc. Natl. Acad. Sci. USA **100** (2003) 9680-9685
50. Bonabeau, E., Dorigo, M., Theraulaz, G., Inspiration for Optimization from Social Insect Behaviour. Nature **406** (2000) 39-42
51. Theraulaz, G., Bonabeau, E., Nicolis, S. C., Sol, R. V., Fourcassi, V., Blanco, S., Fournier, R., Joly, J. L., Fernandez, P., Grimal, A., Dalle, P., Deneubourg, J. L., Spatial Patterns in Ant Colonies. Proc. Natl. Acad. Sci. USA **99** (2002) 9645-9649
52. Kriger, M. J., Billeter, J. B., Keller, L., Ant-Like Task Allocation and Recruitment in Cooperative Robots. Nature **406** (2000) 992-995
53. Helbing, D., Keltsch, J., Molnr, P., Modelling the Evolution of Human Trail Systems, Nature **388** (1997) 47-50
54. Glade, N., Demongeot, J., Tabony, J., Microtubule Self-Organisation by Reaction-Diffusion Processes Causes Collective Transport and Organisation of Cellular Particles. BMC Cell Biol. **5** (2004) epub
55. Rennard, J.-P., Implementation of Logical Functions in the Game of Life. In: Adamatzky, A. (ed.): Collision-Based Computing. London:Springer (2002) 491-512
56. Rendell, P., Turing Universality of the Game of Life. In: Adamatzky, A. (ed.): Collision-based computing. Springer-Verlag (2002) 513-539
57. Dogterom, M., Leibler, S., Physical Aspects of the Growth and Regulation of Microtubule Structures. Phys. Rev. Lett. **70** (1993) 1347-1350
58. Odde, D. J., Estimation of the Diffusion-Limited Rate of Microtubule Assembly. Biophys. J. **73** (1997) 88-96
59. Tabony, J., Glade, N., Demongeot, D., Papaseit, C., Biological Self-Organization by Way of Microtubule Reaction-Diffusion Processes. Langmuir **18** (2002) 7196-7207
60. Tabony, J., Morphological Bifurcations Involving Reaction-Diffusion Processes During Microtubule Formation. Science **264** (1994) 245-248
61. Fygenson, D. K., Marko, J. F., Libchaber, A., Mechanics of Microtubule-Based Membrane Extension. Phys. Rev. Lett. **79** (1997) 4597-4500
62. Kong, C. R., Bursac, N., Tung, L., Mechanoelectrical Excitation by Fluid Jets in Monolayers of Cultured Cardiac Myocytes. J. Appl. Physiol. **98** (2005) 2328-2336
63. Salmon, E. D., Saxton, W. M., Leslie, R. J., Karow, M. L., McIntosh J. R., Diffusion Coefficient of Fluorescein-Labeled Tubulin in the Cytoplasm of Embryonic Cells of a Sea Urchin: Video Image Analysis of Fluorescence Redistribution After Photobleaching. J. Cell. Biol. **99** (1984) 2157-2164
64. Kao, H. P., Abney, J. R., Verkman, A. S., Determinants of the Translational Motility of a Small Solute in Cell Cytoplasm. J. Cell. Biol. **120** (1993) 175-184

65. Adamatzky, A., Programming Reaction-Diffusion Processors. In: Bantre, J. P. et al (eds.): UPP 2004. Lecture Notes in Computer Sciences, Vol. 3566. Springer-Verlag, Berlin Heidelberg (2005) 31-45
66. Adamatzky, A., de Lacy Costello, B., Melhuish, C., Ratcliffe, N., Experimental implementation of Mobile Robot Taxis with Onboard Belousov-Zhabotinsky Chemical Medium. Mat. Sc. Engineering C. **24** (2004) 541-548
67. Tsuda, S., Zauner, K. P., Gunji, Y. P., Robot Control: From Silicon Circuitry to Cells. In: Ijspeert, A. J., Masuzawa, T., Kusumoto, S. (eds.): Biologically Inspired Approaches to Advanced Information Technology, Second International Workshop, BioADIT 2006, Osaka, Japan, Proceedings. Springer (2006) 20-32
68. Adamatzky, A., Collision-Based Computing in Belousov-Zhabotinsky Medium. Chaos, Solitons & Fractals **21** (2004) 1259-1264
69. Platt, M., Muthukrishnan, G., O Hancock, W., Williams, M. E., Millimeter Scale Alignment of Magnetic Nanoparticle Functionalized Microtubules in Magnetic Fields. J. Am. Chem. Soc. **127** (2005) 15686-15687

Perfusion anodophile biofilm electrodes and their potential for computing

John Greenman[1,2], Ioannis Ieropoulos[1,2] and Chris Melhuish[2]

[1] School of Biomedical Sciences, Faculty of Applied Sciences, University of the West of England, Bristol, BS16 1QY, UK
John.Greenman@uwe.ac.uk
http://science.uwe.ac.uk/StaffPages/JG
[2] Bristol Robotics Laboratory, University of Bristol and University of the West of England, Bristol Business Park, Bristol, BS16 1QY, UK

Abstract. This paper presents a theoretical approach to biological computing, using biofilm electrodes by illustrating a simplified Pavlovian learning model. The theory behind this approach was based on empirical data produced from a prototype version of these units, which illustrated high stability. The implementation of this system into the Pavlovian learning model, is one example and possibly a first step in illustrating, and at the same time discovering its potential as a computing processor.

1 Introduction

We have recently described what we now term as perfusion anodophile biofilm electrodes (PABES) and have demonstrated their potential for basic binary type computing [1]. In particular we have shown that the biofilm system can maintain a dynamic steady state under one (of many possible) particular set of physicochemical conditions and then switch to a new steady state in response to changes in one of the parameters of the physicochemical environment (set by the operator) leading to a new condition. We also proposed how interconnecting units might be configured into logic gates (AND , OR , XOR) in order to perform basic binary logic operations [1]. Connections could be made via electrical or fluidic links. A schematic representation of the PABE unit and the type of physico-chemical changes it will respond to is shown in Greenman et al [1], and the mathematical modeling for these units is described below.

1.1 Theoretical modeling of PABES

Some aspects of growth can be modeled using mechanistic equations of the type described by Monod [2]. Anodophilic biofilm cells exist in a non accumulative steady state. Even though the numbers retained in the biofilm remain constant, cells are nevertheless produced and shed into the perfusate. The growth rate (speed) and yield of new cells (biofilm + perfusate cells) can be modelled using Monods equations:

$$\mu = \frac{\mu_{max} \times S}{K_S + S} \tag{1}$$

where: μ = specific growth rate (h^{-1})
μ_{max} = maximum specific growth rate (limiting substrate is supplied in excess)
S = substrate concentration
K_s is Monod's constant = is the substrate utilisation constant, which is numerically equal to the substrate concentration when $\mu = \frac{1}{2}\mu_{max}$

$$Y = \frac{dX}{dS} \tag{2}$$

where: Y = specific growth yield constant
dX = change in concentration of cells
dS = change in the concentration of substrate

For PABES in optimum substrate-limiting conditions the growth rate in the biofilm is given by:

$$\mu\left(h^{-1}\right) = \frac{\text{production rate of cells in perfusate (units } = \text{cfu/mL} \times \text{mlh}^{-1})}{\text{biofilm population (units } = \text{cfu/biofilm)}} \tag{3}$$

Substrate utilisation rate: For PABEs in well-mixed continuous flow:

Substrate in (QSo)= substrate out (QS)+ substrate utilisation K_1ES/Ks+ S

$$\Rightarrow \frac{K_1 ES}{Ks + S} \tag{4}$$

where: Q = volumetric flow rate of bulk fluid
So = initial substrate concentration
S = final substrate concentration
K_1 = rate of reaction (where Vmax = $K_1 E$)
Ks = Monod's (or Michaelis-Menten) constant

E = cell (or enzyme) concentration

In terms of electrical output (Coulombs): Substrate breaks down and is transformed into carbon dioxide, protons and electrons. The theoretical maximum yield from 1 molecule of acetate is 8 electrons, according to the half-rate equation shown below:

$$CH_3COOH + 2H_2O \rightleftharpoons 2CO_2 + 8H^+ + 8e^-$$

In practice, electron abstraction efficiency can be as high as 96% [3,4]. Using the aforementioned equations it would be possible to theoretically model some aspects of individual units.

1.2 A simple learning problem to solve (based on Pavlovian associative learning)

We now propose to demonstrate (in a Gedanken model) how PABES may be configured to produce associated learning, such as that described by Pavlov (1927) and shown in simplified manner in Fig. 1 [5]. Starting with the simplest case we can think of (smell of food, associated sound of bell, switch-on of salivation), we can reduce the original behavioural response to the action of single series of neurones (see Fig. 2). The challenge is to simulate the effects of smell (which can always switch on salivation), bell (which alone does not switch on salivation) and association of the two (learning cycle) such that the bell alone can now trigger the response. Features of this learning function include the number of associations, their duration, the cycle or frequency of repeats and the final loss of associated reflex following a long period of no association.

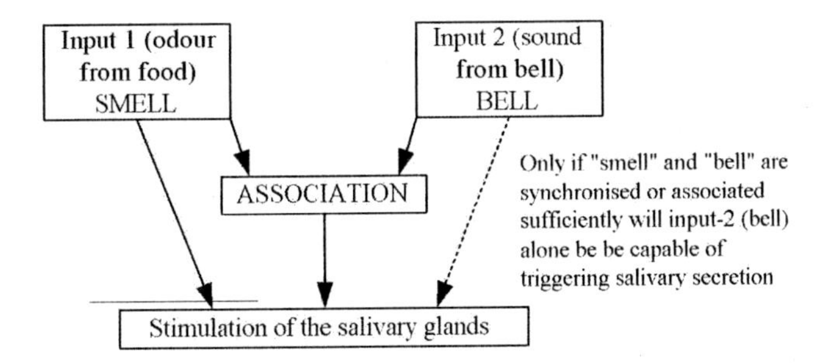

Fig. 1. Simplified representation of Pavlovian feeding reflex occurring in mammals (e.g. dog).

Figure 2 below shows a highly simplified neuronal model of the natural system, and in reality thousands of neurons are likely to be involved. This is an "Ockham's razor" minimum representation; the smallest number that could explain the phenomena. Nevertheless, it presents the case for study with the proposed PABEs. The neuron produces an output along its axon (firing pattern) if the collective effect of its inputs reaches a threshold. The axon from one neuron can influence the dendrites of another neuron via junctions (synapses). Some synapses may generate a positive (stimulatory) effect in the dendrite encouraging its neuron to fire, whilst others produce a negative (inhibitory) effect reducing the propensity to fire. A single neuron may receive inputs from 1000's of synapses and the total number of synapses in the human brain has been estimated to be of the order of 10^{15} [6]. Learning and memory reside within the circuits formed by the interconnections between the neurons (i.e. at the synapses) [6].

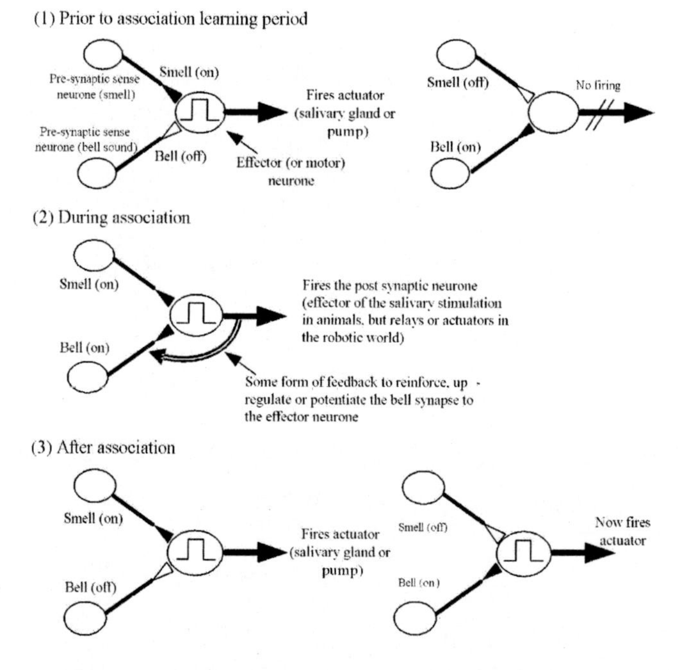

Fig. 2. Translation of Pavlovian response as simplified neuronal connections using the minimal number of neurones (3). The closed triangles indicate active synapses whilst the open triangles indicate inactive synapses.

1.3 Memory based on substrate storage in a dynamic dilution reservoir

The fluid input into a PABE unit is fed from a reservoir of perfusate containing minimal maintenance medium with or without substrate, depending on the reservoir inputs (see Fig. 3). The reservoir is a small continuous flow, stirred tank reactor to effect changes in the profile of substrate concentration with time. The dilution rate (D) is given by:

$$D = \frac{f}{V}\ [h^{-1}] \tag{5}$$

where: f = totality of liquid substrates flow rate (ml/h)
V = volume of the liquid in the vessel (typically measured in ml.).

The concentration profile measured at the output depends on the substrate concentration (e.g. acetate), which varies depending on the ratio of substrate stock or buffer diluent that are mixed together upon entering the reservoir. A simple case would be a quick pulse of strong substrate. The concentration profile thereafter can be both empirically measured and theoretically modelled as an exponential decay [7]. The rate or frequency of pulse additions into the memory vessel can now be calculated that would be needed to maintain the output concentration above or below threshold limits to maintain or activate PABES.

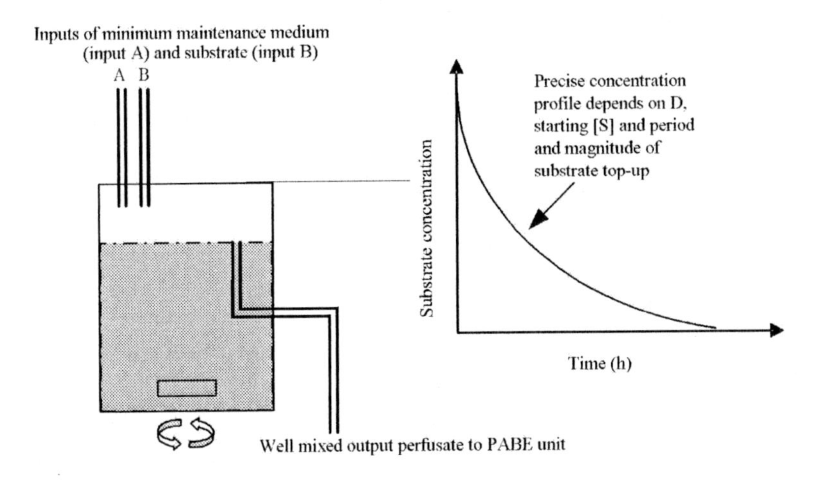

Fig. 3. A chemical reservoir of substrate may be used to hold memory

Increasing D of the maintenance substrate flow rate (without further addition of substrate) will decrease the residence time of the substrate molecules and 'memory' is more rapidly lost, requiring frequent top-up of substrate to keep the 'memory' in store. Decreasing D of the maintenance substrate flow rate will decrease the rate of substrate wash-out and lengthen the period of memory. Other dynamic mixing systems could also be applied for practical purposes to control the profile of substrate.

2 Possible implementation as a solution to the problem

To implement the learning behaviour, the input we call 'smell' is represented by a switch. The switch may operate an actuator that brings substrate (akin to smell molecules) into a PABE unit (Unit-1), which senses the molecules and is switched on (see Fig. 4).

When PABE Unit-1 is ON it then activates Unit-2. This link could be a direct electrical link to Unit-2, whereby the output from Unit-1 changes the electro-potential of Unit-2 from suboptimum to optimum. Alternatively, the output of Unit-1 switches a relay or actuator, which then switches ON a substrate stream into Unit-2. As a result, Unit-2 becomes active switches ON the final pump labeled 'salivary gland'. Thus, whenever 'smell' is ON, it always switches ON the 'salivary gland'.

This is not the case for Unit-3, which is labeled as 'bell'. When the 'bell' switch is ON, it adds weak substrate into the perfusate to partially activate Unit-3. However, this weak activity gives an output that is insufficient to switch ON the substrate supply to Unit-2, and hence this unit is OFF (effectively the 'salivary gland' pump is OFF). Although the output from Unit-3 is insufficient to activate Unit-2, it is nevertheless sufficient to activate one of the two inputs into the AND gate. The implementation of an AND gate using a single PABE unit has already been described [1]. The other input to the AND gate is obtained from the electrical output of PABE Unit-2.

In associative learning, both 'smell' and 'bell' are switched ON at the same time. This means that Unit-2 (giving one input to the AND gate) and Unit-3 (giving the other input to the AND gate) are both ON simultaneously. The AND gate is now switched ON and its output is used to switch ON a strong substrate solution to dose the vessel called 'memory reservoir'. Volume from this reservoir is displaced into Unit-3. If the period of association is sufficient, the concentration of 'memory' substrate will build-up in the reservoir and the output concentration combined with the weak substrate entering Unit-3, are sufficient to produce a high output which will fully switch ON Unit-2.

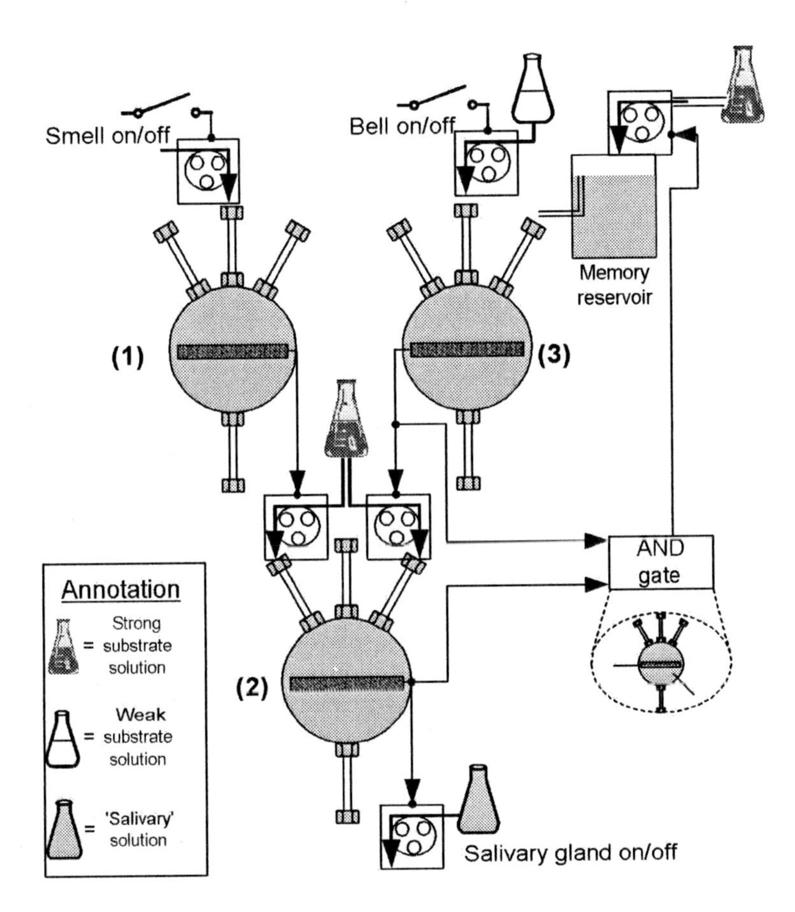

Fig. 4. Implementation using memory, 3 PABE units, plus one unit for the AND gate.

On the next occasion, when the 'bell' alone is switched ON, the combination of output from the weak substrate pump and the 'memory reservoir' output are together sufficient to fully switch ON Unit-3. This is then capable of switching ON Unit-2 with a resultant effect of switching ON salivary secretion.

The decay of 'memory' depends on (a) the dilution rate of the 'memory reservoir' vessel and (b) the frequency of addition of substrate to the 'memory vessel'. The latter depends on co-activation of the AND gate by Units-2 and -3. If 'bell' is not used for a long period of time, the memory may decay to below threshold. This period may be 'set' by the operator by controlling 'memory' dilution rate (i.e. D). Another period of associative learning (smell and bell ON simultaneously) would then be required before 'bell' alone could switch ON the 'salivary' system.

It should be noted that all PABE units, including the 'memory reservoir' throughout the experiment are constantly perfused with maintenance nutrient buffer and are all in continuous readiness (viable steady-state) to respond to changes in conditions.

3 Discussion

The envisaged microbial processing units, are a symbiotic mix between the natural biological cells (the anodophiles) and artificial systems (electrodes, actuators, pumps and chemical solutions) in which the microbial cells individually mediate critical information-processing functions. In the simple example given, the processes are achieved by programming the units to have 2 different types of input (electrical and fluidic) and 3 electrical output states; OFF, ON and INT (intermediate-partially ON) e.g. Unit-3. In this way, whereas ON or OFF are hard-thresholds, which either activate or have no-effect on different parts of the system, the INT output can be used as either. In the Pavlovian learning example given above, the same INT output from Unit-3 denotes low/weak substrate conditions therefore having no effect on the next stage — Unit-2 — and at the same time activates a different part of the system — the AND gate. The processing described is fairly trivial and possibly undervalues the number of states that could be realized and applied to achieve useful computing. Nevertheless this is perhaps a first step in the direction of developing such biological computing systems.

The described PABE model would still be considered to be operating in binary mode, even though the multitude of inputs and states of output may be suggesting otherwise. In other words, the operation is effectively 'learning' by association between a particular state and a conditional stimulus. The PABE units are on the other hand potentially capable

Table 1. Different states of the PABE-physicochemical domain

Parameter (defined and controlled)	Possible parameter increments	Total states
C/E limiting substrate (e.g. acetate)	8	8^1
Types of C/E substrate	8	8^2
Other types of limiting nutrients (e.g. S, P, N, Mg^{2+}, K^+, Fe^{2+} etc.)	8	8^3
Binary or tertiary mixes of selected substrates	8	8^4
Osmolarity (NaCl)	8	8^5
Temperature	8	8^6
pH	8	8^7
Reversible competitive inhibitors	8 types, 8 concentrations	8^9
Flow rate of perfusate medium	8	8^{10}
Electrode overpotential	8	8^{11}
Number of PABE units	8×8	8^{13}

of multiple inputs and, more importantly, multiple outputs of different forms. As described by Greenman et al [1] and partially illustrated by Unit-3 in the Pavlovian learning example, the various (multiple) forms of input and output can also have multiple states, which increase the level of computational complexity that may be achieved. It may be set up to have only one state and associate with multiple stimuli or furthermore to have multiple states associating with multiple stimuli. The totality of input and output states in these units may be advantageously employed to possibly attempt to solve problems that cannot be solved by conventional computers.

A more useful expression of their processing power may lie in solving different types of problems particularly those involving multi-logic processing. Their information-processing virtuosity traces ultimately to the fact that they possess macromolecules, most notably proteins (e.g. substrate transport receptors or membrane transporters), that can recognize specific molecular objects in their environment (e.g. acetate, the main substrate) in a manner that uses shape and depends sensitively on physiochemical context [8]. The number of interacting physicochemical parameters that can be both controlled by the operator and would be expected to interact with the biofilms is huge. For purposes of computing, the number of environmental (milieu) factors that could conceivably be used to encode the input signals is also virtually boundless. Thus, for each of the named parameters listed in table 1, it is likely that 8 incremental states or separations of parameter magnitude can be recognized (resolved). For example, for a particular substrate (e.g. acetate), 8 different concentrations (all rate-limiting) would be expected to be resolvable into 8 different states, each being reproducible and repeatable over time. Since there are many such parameters (Table 1), the total number of discernable yet reproducible/repeatable states (i.e. specified states) could be as many as 8^{13}.

The biofilm cells are the primary processing component, acting on the chemical substrate to produce an electrical output that is directly proportional to their ultimate output, i.e. growth. Proteins (enzymes and receptors) may assume numerous molecular shapes. Within the cell (under physiological conditions) a subset of molecular shapes is favoured. Moreover, under different physiological states the whole cell expresses different genes and gives rise to a different proteomic landscape. This allows the whole cell to fuse the information concerning the physical world into the single output (growth). It does this in a highly complex non-linear manner that would require a large number of conventional processors to simulate. Biological cells are replete with receptors that convert signals representing macrofeatures of the external environment into 'internal' signals that are susceptible to control by meso- and microlevel processing (by enzymes, pathways, protein expression etc.) [8]. Biomolecular architectures are sharply different to silicon designed systems since their complexity is inherent. For making PABES practical, the amount of computational work performed at the meso- or micro levels should be as much as possible since (a) this is thermodynamically favoured (in contrast to macroscopic signals from the electrode onwards, which is thermodynamically costly), and (b) the non-linear nature of complexity that is available at this level. The switch from one physiological state to another in response to even a single parameter change in the physiocochemical environment (real world) is highly complex and non-linear. It is here that powerful context-sensitive input-output computing transformations reside.

The output from a PABE is the totality of activity from a collection of about $10^7 - 10^8$ microbial cells. Although each cell is exposed to almost identical physicochemical conditions their growth cycle and division points are asynchronous. Although there is indeterminancy at the level of any one cell, this gives way to predictability at the level of the whole biofilm and a system with self-organising dynamics (the PABE has the potential to 'hold' thousands of different recognition-action, input-output transforms) may be advantageous to perform more complex operations than systems with programmable architectures. PABES can be networked by using wires or liquid flows and synapse-like inputs strenthened or inhibited by changing physicochemical environment by either electrical or fluid link. Changes can be induced in one part of the network by designing connections, and modulating these connections through a learning process.

A systematic machine approach whereby PABES are integrated with conventional *in silico* devices could be explored and used to produce an interesting processing engine for solving certain types of computing tasks. Sculpting the desired functionality by coding of the input signals and gaining knowledge of the operational conditions that fuse the main inputs into the main output (growth) is the main challenge. Therefore, the ultimate goal of our work is to create a wide repertoire of high-complexity functions for implementing input-output transforms that cannot easily be done by current digital computers with programmable architectures. In our system a large portion of the complexity and initial processing is natural and endogenous to the living cell. There is less need for an outside programmer since programmable functionality can be molded through designed or natural adaptive procedures, maintenance functions, set parameters and feedback.

Consider a change from null to any other condition and then a change back to the original null state. During this process, the cells performs 1000's of automatic back-calculations or adjustments (by default) in order to (fairly) rapidly return to the null-condition, so that the operator can set the next computational 'problem'. This is, at the moment, non-transparent to the human operator and accessing this micro/meso level of complexity, may succeed in achieving a high level of computing power.

In addition to the high level of computational capability, the transformation of substrate(s) into electrons allows the possibility for the c electrical output of many units to be sufficiently strong to drive low-power actuators (or co-processors) giving the system energy autonomy, which may be ideally utilised for some applications, including small scale robots.

References

1. Greenman, J., Ieropoulos, I., McKenzie, C. and Melhuish, C., Microbial Computing using Geobacter biofilm electrodes: output stability and consistency. Int. J. Unconventional Computing (2007) (to appear).
2. Monod, J., La technique de culture continue: therie and applications. Ann. Inst. Pasteur **79** (1978) 390–410.
3. Bond, D.R., Lovley, D.R., Electricity Production by Geobacter sulfurreducens Attached to Electrodes. Appl. Environ. Microbiol. **69**(3) (2003) 1548–1555.

4. Ieropoulos, I., Greenman, J., Melhuish, C. and Hart, J., Comparison of three different types of microbial fuel cell. Enzyme and Microbial Technology **37**(2) (2005) 238–245.
5. Pavlov, I. P., Conditioned Reflexes. (Translated by G. V. Anrep) London: Oxford (1927).
6. Murre, J.M.J. and Sturdy, D.P.F., The connectivity of the brain: multi-level quantitative analysis. Biol Cybernet **73** (1995) 529–545.
7. Borzani, W. and Vairo, M.L.R., Observations of continuous culture responses to additions of inhibitors. Biotechnology and Bioengineering, **15**(2) (1973) 299–308.
8. Conrad, M. and Zauner, K.,-P.: Conformation-based computing: A rationale and a recipe, In: Sienko T., Adamatzky A., Rambidi, N.G. and Conrad, M. (Eds.), Molecular Computing, MIT Press, Cambridge, Massachusetts (2003).

Implementations of a model of physical sorting

Niall Murphy[1], Thomas J. Naughton[1], Damien Woods[2], Beverley Henley[3], Kieran McDermott[3], Elaine Duffy[4], Peter J. M. van der Burgt[4], and Niamh Woods[4]

[1] Dept. of Computer Science, National University of Ireland, Maynooth, Ireland
nmurphy@cs.nuim.ie, tom.naughton@nuim.ie
[2] Boole Centre for Research in Informatics, School of Mathematics, University College Cork, Ireland
d.woods@bcri.ucc.ie
[3] Dept. of Anatomy, BioSciences Institute, University College Cork, Ireland
b.henley@ucc.ie
[4] Dept. Experimental Physics, National University of Ireland, Maynooth, Ireland
peter.vanderburgt@nuim.ie

Abstract. We define a computational model of physical devices that have a parallel atomic operation that transforms their input, an unordered list, in such a way that their output, the sorted list, can be sequentially read off in linear time. We show that several commonly-used scientific laboratory techniques (from biology, chemistry, and physics) are instances of the model and we provide experimental implementations.

1 Introduction

There has been interest in identifying, analysing, and utilising computations performed in nature [1,7,8,10,11,16,19,21,23,26,28], in particular where they appear to offer interesting resource trade-offs when compared with the best-known sequential (e.g. Turing machine) equivalent. In this paper we present a special-purpose model of computation that falls into that category. Other natural sorting algorithms have been proposed in the literature [2,3,24].

Natural scientists routinely separate millions of particles based on their physical characteristics. For example, biologists separate different lengths of DNA using gel electrophoresis [25], chemists separate chemicals by using chromatography [22], and physicists separate particles based on their mass-to-charge ratio using mass spectrometry [15]. The common idea behind these techniques is that some physical force affects objects to an amount that is proportional to some physical property of the objects. We use this idea to sort objects and provide a special-purpose model of computation that describes the method idea formally. In this paper we present five instances of the model that are routinely used for ordering physical objects but, to our knowledge, all but one have never before been proposed for sorting lists of numbers. We also provide four physical implementations utilised frequently by scientists in the fields of chemistry, biology, and physics. In these fields, particles of diameter 10^{-6} meters and below are sorted. This suggests that the model can be used for massively parallel computations.

The special-purpose model works as follows. We encode the list of numbers (input vector) in terms of one physical property of a chosen class of particle, such that an easily realisable known physical force will affect proportionally each particle based on its value for that property. The one-dimensional (1D) input vector is transformed to a two-dimensional (2D) matrix via a constant-time atomic operation which represents the action of the force. This 2D matrix representation admits a simple linear time algorithm that produces a (stable) sort of the original 1D input list.

In Section 2 we introduce the Model of Physical Sorting (which we simply refer to as the Model). We define the (general) Model in Section 2.1 and the Restricted Model in Section 2.2. One of the interesting aspects of the Model is that several implementations of it already exist; in Section 3 we describe five instances and present experimental results for four that are used as real-world sorting methods. Some of these instances are not instances of the Model but of the Restricted Model. In Section 3.4 we show how a pre-existing instance of the Restricted Model is generalised to become an instance of this Model. Section 4 concludes the paper.

2 Model of physical sorting

In this section we introduce the Model of Physical Sorting and the Restricted Model of Physical Sorting. Their instances take as input a list $L = (l_1, l_2, \ldots, l_n)$ and compute the *stable sorting* [17] of the list.

Definition 1 (Stable sort).

A sort is stable if and only if sorted elements with the same value retain their original order. More precisely, a stable sorting is a permutation $(p(1), p(2), \ldots, p(n))$ *of the indices* $\{1, 2, \ldots, n\}$ *that puts the list elements in non-decreasing order, such that* $l_{p(1)} \leqslant l_{p(2)} \leqslant \cdots \leqslant l_{p(n)}$ *and that* $p(i) < p(j)$ *whenever* $l_{p(i)} = l_{p(j)}$ *and* $i < j$.

Not all sorting algorithms are stable; we give some counterexamples. Clearly any sorting algorithm that does not preserve the original relative ordering of equal values in the input is not stable. A sorting algorithm that relies on each element of its inputs being distinct to ensure stability is not stable. A sorting algorithm that outputs only an ordered list of the input elements (rather than indices) is not necessarily stable.

2.1 The model

Let $\mathbb{N} = \{1, 2, 3, \ldots\}$. Before formally describing the computation of the Model we give an informal description. The input list is transformed to a 2D matrix that has a number of rows equal to the input list length and a number of columns linear in the maximum allowable input value. The matrix is zero everywhere except where it is populated by the elements of the input list, whose row position in the matrix is their index in the input and whose column position is proportional to their value. The values in the matrix are then read sequentially, column by column, and the row index of each nonzero value is appended to an output list. This output list of indices is a stable sorting of the input list.

Definition 2 (Model of Physical Sorting). *A Model of Physical Sorting is a triple* $S = (m, a, b) \in \mathbb{N} \times \mathbb{N} \times \mathbb{N}$, *where* m *is an upper bound on the values to be sorted and* a, b *are scaling constants.*

The Model acts on a list $L = (l_1, l_2, \ldots, l_n)$ where $l_i \in \{1, \ldots, m\}$ and m is some constant that is independent of n. Given such a list L and a Model of Physical Sorting S we define a $n \times (am + b)$ matrix G with elements

$$G_{i,j} = \begin{cases} l_i & \text{if } j = al_i + b \\ 0 & \text{otherwise} \end{cases} \tag{1}$$

An example G for given S and L is shown in Fig. 1.

Definition 3 (Physical Sorting computation). *A Physical Sorting computation is a function* $c : \{1, \ldots, m\}^n \rightarrow \{1, \ldots, n\}^n$ *that maps a list* L *of values to a sorted list of indices*

$$c(l_1, l_2, \ldots, l_n) = (k_1, k_2, \ldots, k_n) \tag{2}$$

where l_{k_p} *is the* p^{th} *non-zero element of* G *and the elements of* G *are assumed to be ordered first by column and then by row.*

	1	2	3	4	5	6	$7 = am + b$
1			1				
2							3
3					2		
4			1				
5			1				
$n = 6$							3

Fig. 1. Graphical illustration of the matrix G for example model $S = (m, a, b) = (3, 2, 1)$ and for example input list $L = (1, 3, 2, 1, 1, 3)$.

Remark 1. A Physical Sorting computation returns a stable sorting of its input: k_1 is the index of the first value in the stable sorting of L, k_2 is the index of the second value, and so on.

Remark 2. We assume that a Physical Sorting computation is computed in at most $(am + b)n + 1 = O(n)$ timesteps. The creation of matrix G takes one timestep and obtaining the indices of the nonzero values in G takes one timestep per element of G.

Each of the physical instances of our Model of Physical Sorting that follow are consistent with Remarks 1 and 2; the matrix G is generated in a single parallel timestep and the Physical Sorting computation takes linear time to output a stable sorting.

An interesting feature of the algorithm is the fact that it has a parallel part followed a sequential part. One could ask that the entire algorithm be either entirely sequential or entirely parallel, with a respective increase or decrease in time complexity. However, neither of these scenarios correspond to the way in which physical instances (see Section 3) are actually performed in the laboratory.

2.2 Restricted model

It is possible to restrict some physical instances of the Model in Section 3 to become instances of the Restricted Model. This restriction is achieved by removing the abilities to track indices and deal with repeated elements.

Definition 4 (Restricted Model of Physical Sorting). *A Restricted Model of Physical Sorting is a triple* $S = (m, a, b) \in \mathbb{N} \times \mathbb{N} \times \mathbb{N}$*, where* m *is an upper bound on the values to be sorted and* a, b *are scaling constants.*

The Restricted Model acts on a multiset $T = \{t_1, t_2, \ldots, t_n\}$, $t_i \in \mathbb{N}$ and m is some constant that is independent of n. Given such a multiset T and a Restricted Model of Physical Sorting S we define the vector V of length $am + b$. As before a, b are scaling constants and $m = \max(T)$. The vector V has elements

$$V_j = \begin{cases} t_i & \text{if } j = at_i + b \\ 0 & \text{otherwise} \end{cases} \tag{3}$$

Definition 5 (Restricted Physical Sorting computation). *A Restricted Physical Sorting computation is a function* c *that maps a set* $T = \{t_1, t_2, \ldots, t_n\}, t_i \in \mathbb{N}$ *to a list*

$$c(T) = (t_{k_1}, t_{k_2}, \ldots, t_{k_n}) \tag{4}$$

where t_{k_p} *is the* p^{th} *non-zero element of* V.

It is not difficult to see that $c(T)$ is a list of strictly increasing values, that is $t_{k_i} < t_{k_{i+1}}$ for all $i \in \{1, 2, \ldots, n-1\}$.

The Restricted Model computes a non-stable sorting of the input multiset.

Remark 3. The input to a Restricted Model of Physical Sorting is a multiset, however the output vector does not contain any duplicated elements. Also the output of a Restricted Physical Sorting computation is a sorted list of the input elements, no index information is available and so is not necessarily a stable sort.

Remark 4. We assume that a Restricted Physical Sorting computation is computed in at most $(am + b) + 1 = O(1)$ timesteps. The creation of vector V takes one timestep and obtaining the sorted list in V takes one timestep per element of V.

3 Physical instances of the model

In this section we give five example instances of either the Model, the Restricted Model that arise in commonly-used scientific laboratory techniques, gel electrophoresis, chromatography, the dispersion of light, optical tweezers, and mass spectrometry. A brief introduction is given to each instance and we explain how it is used to sort. Most of the examples are instances of both the Model and the Restricted Model and we show how the examples compute in a way that is consistent with the models. In all but one case (optical tweezers) we present an experimental implementation of the physical instance.

With Gel Sort and Rainbow Sort the matrix produced is the mirror image of the other examples; larger input elements have a smaller column index in the matrix while smaller input elements have a large column index. Reading the matrix of sorted values in the normal way yields an output list of non-increasing order. To get an output list of non-decreasing order we read the matrix starting with the largest column index.

3.1 Gel sort

Gel electrophoresis [25] is a fundamental tool of molecular biologists and is a standard technique for separating large molecules (such as DNA and RNA) by length. It utilises the differential movement of molecules of different sizes in a gel of a given density.

Description The process of gel electrophoresis is illustrated in Fig. 2 and occurs as follows. Samples of DNA molecules are placed into wells at one end of a rectangle of agarose gel. The wells are separated from each other at different spatial locations along a straight line. The gel is then permeated with a conducting liquid. Electrodes apply a voltage across the gel which provides a force upon the charged molecules causing them to be pulled towards the oppositely charged electrode. Smaller molecules move through the gel more quickly and easily than larger molecules. This difference in velocity orders the molecular samples by number of base pairs.

Sorting Here we briefly describe Gel Sort; using gel electrophoresis for sorting. Given a list L of numbers to be sorted, we encode each element of L as a sample of molecules with a number of base pairs proportional to the element value. Each sample of uniform length molecules is placed (in the same order as in L) in the wells at one end of the gel. A voltage is applied for a time and the molecules move through the gel at a rate inversely proportional to their length. When the voltage is removed the gel is a representation of the matrix G (see Definition 2). We then read off the list of sorted indices by recording the index of each element in order of those which traveled the least and in order of their index. The resulting list is in decreasing order. Restricted Gel Sort is similar to Gel Sort except that all the samples of molecules are placed in the same well.

Instance Viney and Fenton [27] provide an equation that describes the physics of gel electrophoresis,

$$V = K_1 \frac{E}{\varepsilon M^n} - K_2 E\,, \tag{5}$$

where V is the velocity of a molecule of molecular mass M in an electric field E, where the ratio between the pore size and the typical size of the molecules is given by $0 < n \leqslant 1$, and where ε is the permittivity of the conducting liquid. The constants K_1 and K_2 represent quantities such as the length of the gel, and the charge per unit length of the molecule [27].

To get distance s we apply $V = s/t$ where t is time, giving

$$s = K_1 \frac{Et}{\varepsilon M^n} - K_2 Et\,.$$

We refer to sorting using gel electrophoresis as Gel Sort. For an instance of Gel Sort we choose appropriate values for $K_1, K_2, E, t, \varepsilon \in \mathbb{R}$ such that $k_1 = (K_1 Et/\varepsilon) \in \mathbb{N}$ and $k_2 = (K_2 Et) \in \mathbb{N}$. We also let $n = 1$ which gives

$$s = k_1 M^{-1} - k_2\,. \tag{6}$$

Equation (6) satisfies Equation (1) if we let $S = (m, k_1, k_2)$ where $m \in \mathbb{N}$ is the smallest length of DNA or RNA to be sorted.

Given a list L to be sorted, we encode each list value as a sample of molecules of proportional length. Each sample of molecules is then placed in an individual well, in the same order of the list to be sorted L. After the gel is run, it is a representation of the matrix G. To read the list of indices corresponding to a stable sort, we sequentially record the indices of the samples beginning with those that traveled least. Thus Gel Sort implements arbitrary computations of the Model.

Given a multiset T to be sorted, we encode each multiset value as a sample of molecules of proportional length. If we place all molecules in one well, after the gel is run it is a representation of the vector V. We read the sorted list by recording the length of each sample beginning with those that traveled least. Thus Restricted Gel Sort is an instance of the Restricted Model of Physical Sorting.

Implementation We have implemented an instance of Gel Sort and Restricted Gel Sort. The elements of the unordered list of numbers to sort L = (550, 162, 650, 200,550, 350, 323, 550) are encoded as DNA strings with a number of base pairs proportional to their value. These values of DNA are placed in individual wells in a 1% agarose gel in the order that they appeared in the list L. We also place a sample of each in a single well to produce a non-stable sort for comparison. The gel is run for some time and the result is seen in columns 1 – 8 in Fig. 2. We then read off the list of sorted indices by recording the index of each element in order beginning with those which traveled the least and in order of their index. This yields the list of indices $c(L)$ =(2,4,7,6,1,5,8,3) which yields the sorted list of elements (162, 200, 323, 350, 550, 550, 550, 650).

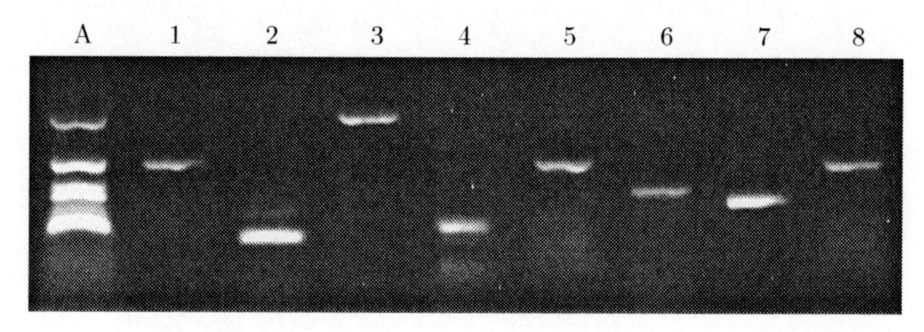

Fig. 2. Electrophoresis of DNA molecules in a 1% agarose gel. Lane A is a non-stable sort which contains strings of DNA with the same number of base pairs as all of those in lanes 1 to 8. In lanes 1 to 8 the DNA molecule lengths are respectively 550, 162, 650, 200, 550, 350, 323 and 550 base pairs.

We have also performed a Restricted Gel Sort that is an instance of the Restricted Model of Physical Sorting (see Definition 5). We sort the set of numbers $T = (550, 162, 650, 200, 550, 350, 323, 550)$ but this time they are all placed in a single well in the agarose gel. The result which is seen in column A of Fig. 2 is $c(T) = (162, 200, 323, 350, 550, 650)$. This is a non-stable sorting of the list L as the multiple instances of the element 550 were lost in the result.

3.2 Optomechanical sort

The movement of small transparent particles by light alone is an effect most commonly employed in optical tweezers [5] for biologists to manipulate micro-scale objects. Several methods of ordering particles using this technology have been proposed [9, 14]. We, however, propose a novel method that is an instance of the model of Physical Sorting. We do not provide a implementation of Optomechanical Sort or an instance of the Restricted Model.

Description It is known that transparent objects experience a force when a beam of light passes through them [4]. This force is caused by the beam's path being refracted by the object. A change in light beam direction causes a change in the beam's momentum, and momentum is only conserved if there is an equal but opposite change of momentum for the object. This momentum change has a component in the same direction as the direction of the beam and a component in the direction of the increasing intensity gradient of the beam (the gradient force, F_{grad}).

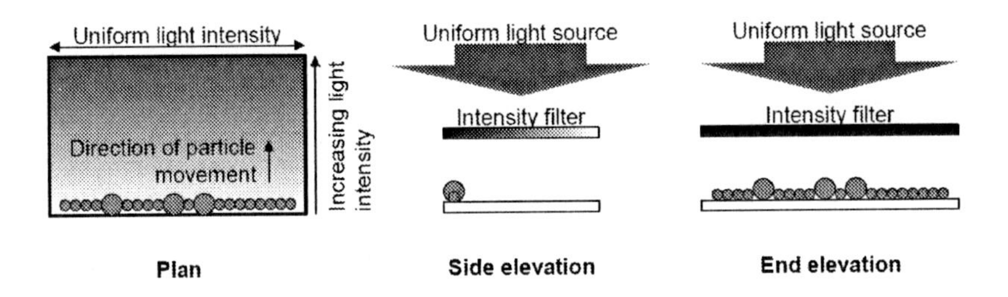

Fig. 3. The initial configuration of Optomechanical Sort. The circles represent the particles to be sorted.

Sorting We propose the use of optical tweezers technology to sort objects and we refer to this sort as Optomechanical Sort. In Optomechanical Sort, all of the input objects are arranged in a straight line in a medium (e.g. water). There is a barrier that prevents the objects from moving in the direction of the beam. A light source, constant in time, and with a strictly increasing intensity gradient perpendicular to the axis of the input objects is applied (see Fig. 3). This intensity gradient is achieved by modulating a uniform light field with an intensity filter variable in one direction only. The objects with a larger volume move more quickly in the direction of increasing intensity than those of a smaller volume. This movement separates the objects according to their volumes.

Instance We proceed by using the equations for objects smaller than the wavelength of the light beam. According to Ashkin [6] the equation to calculate the force in the direction of the gradient on the particles is

$$F_{\text{grad}} = -\frac{n_b^3 V}{2}\left(\frac{m^2 - 1}{m^2 - 2}\right)\nabla E^2$$

where n_b is the refractive index of the medium, m is the refractive index of the particles divided by the index of the medium, V is the volume of the particles and ∇E^2 is the change in beam density over the particle.

For each instance of Optomechanical Sort we let $n_b, m, \nabla E^2$ be constants such that

$$F_{\text{grad}} = k_1 V \tag{7}$$

where $k_1 \in \mathbb{N}$, holds. Equation (7) satisfies Equation (1) with $S = (m, k_1, 0)$ where m is the maximum particle volume for the specific material and medium. The sort is stable as we obtain a list of indices by reading the index of each particle in the order of least distance traveled and since the particles move in parallel lines. Thus Optomechanical Sort is an instance of the Model.

3.3 Chromatography sort

Chromatography is a collection of many different procedures in analytical chemistry [20] which behave similarly (e.g. gas chromatography, liquid chromatography, ion exchange chromatography, affinity chromatography, thin layer chromatography). It is commonly used to separate the components in a mixture.

Description Chromatography separates the input chemicals (analytes) over time in two media; the mobile phase and the stationary phase. The mobile phase is a solvent for the analytes and filters through the stationary phase. The stationary phase resists the movement of the analytes to different degrees based on their chemical properties. This causes the analytes to separate over time.

Sorting We refer to the use of chromatography to sort substances by their average velocity through the stationary phase as Chromatography Sort. We proceed assuming known relative velocities for analytes in our apparatus. The apparatus is either wide enough to accommodate many analytes side by side or is made of several identical setups which allow side by side comparisons.

Given a list L of numbers to be sorted, we encode each element of L as a sample of analyte with a relative velocity proportional to the element value. Each analyte is placed in the chromatography apparatus in the same order as in L. When the process commences the analytes move along the stationary medium at a rate proportional to its relative velocity. When the process is halted the apparatus is a representation of the matrix G (from Equation (1)). We then read off the list of sorted indices by recording the index of each element in order of those which traveled the most and in order of their index (position in L).

Restricted Chromatography Sort is similar to Chromatography Sort except that all analytes are mixed together and placed in the apparatus as one sample.

Instance We use standard equations from analytical chemistry [22] to calculate the distance traveled by an analyte in a particular mobile phase and stationary phase.

Given the time t_m for the mobile phase to travel distance L_m, the average velocity $\bar{u}_m$ of the mobile phase in the stationary phase and the capacity factor k of the analyte, Poole and Schuette [22] provide

$$t_R = \frac{L_m}{\bar{u}_m}(1+k)$$

to find the time t_R that it takes the analyte to travel the distance L_m. They also provide

$$k = \frac{t_R - t_m}{t_m}$$

to find the value of k. By substitution we find

$$L_m = \bar{u}_m t_m \,.$$

It follows that an analyte moving at an average velocity of $\bar{u}_a \leqslant \bar{u}_m$ will in time t_m travel a proportional distance $L_a \leqslant L_m$, that is

$$L_a = \bar{u}_a t_m \,. \tag{8}$$

We refer to the use of chromatography to sort substances by their average velocity $\bar{u}_a$ through the stationary phase as Chromatography Sort. If we provide an instance of the Model with the triple $S = (m, t_m, 0)$ where $m = \bar{u}_m$ is the average velocity of the mobile phase in the stationary phase, it is clear that Equation (8) satisfies Equation (1).

Also, we ensure that Chromatography Sort is stable by running each analyte to be sorted side by side, or on a separate but identical, apparatus. After the run, the apparatus is a representation of the matrix G. The final indices are recorded in order of analytes that traveled the least distance and in the case of analytes traveling the same distance, in the order of the indices starting with the smallest. Thus Chromatography Sort is an instance of the model.

In an instance of Restricted Chromatography Sort all analytes are mixed together and placed in the apparatus as one sample. After the run the apparatus is a representation of the vector V. The sorted list is read by recording the order of the analytes staring with those that traveled the least distance. Restricted Chromatography Sort is an instance of the Restricted Model.

Implementation We have performed an implementation of Chromatography Sort with a list of numbers $L = (3, 1, 2, 3, 2, 1)$. The chemicals involved were domestic food dyes and their average velocities were measured with a water mobile phase and thin layer stationary phase. We assigned each element in L a chemical directly proportional to its average velocity with a water mobile phase and thin layer stationary phase. In this case the assignment was 1 to blue, 2 to red, and 3 to yellow.

The result is seen in Fig. 4. The final indices are read off in order of

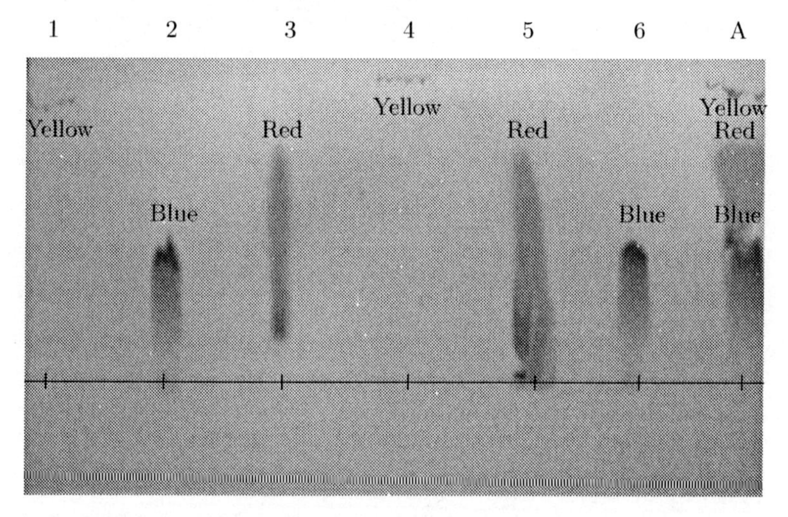

Fig. 4. Chromatography of household food dye in water on a thin layer plate. Lane A is a non-stable sort which contains each of the dyes in lanes 1 to 6. In lanes 1 to 6 the average speeds are, from left to right 0.00003ms^{-1}, 0.00001ms^{-1}, 0.00002ms^{-1}, 0.00003ms^{-1}, 0.00002ms^{-1}, 0.00001ms^{-1}.

least distance traveled by the analytes giving the list $c(L) = (2,6,3,5,1,4)$. These indices yield the stable sorting of L with duplicates in the same order that they appeared in the input. We also performed an instance of the Restricted Model, the output of which is seen in Fig. 4 column A. The input set was $T = (3, 1, 2, 3, 2, 1)$ and the numbers were encoded using the same scheme. The sorted output is $c(T) = (0.00001\text{ms}^{-1}$,

$0.00002\text{ms}^{-1}, 0.00003\text{ms}^{-1}$) which does not record the multiple instances of set elements. The result is a non-stable sorting of the set T.

3.4 Rainbow sort

Rainbow Sort was first described by Schultes [24] as an unstable sort but by a simple generalisation it becomes an instance of the Model.

Description Rainbow Sort utilises the phenomenon of dispersion, where light beams of longer wavelengths are refracted to a lesser degree than beams of a shorter wavelength. Dispersion occurs where there is a change of refractive index in the media (such as an interface between air and glass) in the path that the light beam travels.

Sorting In Rainbow Sort, as described by Schultes [24] each element of a list L is encoded as a distinct wavelength proportional to its value. A beam of light containing only the wavelengths to be sorted is passed through a prism. The component wavelengths are refracted at different angles and so emerge from the prism as separate beams and in an order dictated by their wavelength (see Fig. 5). A light measurement device is positioned to sequentially read the ordered component beams. Schultes considers the input encoding (which takes linear time) in his complexity analysis which differs from our constant time analysis. This is an unstable sort as it does not output repeated elements that were in the input and as defined does not return a list of indices and so is not necessarily stable.

Schultes provides a possible technique to sort lists with repeated elements with Rainbow Sort [24]. We suggest our own method that follows from the Model of Physical Sorting called Generalised Rainbow Sort, which is similar to Rainbow Sort except that it utilises the full geometry of the prism and is an instance of the Model. It also returns the indices of the sorted list elements, thus guaranteeing stability. Each element of the list L is encoded as a beam of light of a distinct wavelength proportional to its value. Each beam is then passed through the prism at a different depth in the prism, as shown in Fig. 5. We then read off the list of sorted indices by recording the index of each refracted beam in order of those which where refracted the most and were there are multiple beams refracted to the same degree, in order of their index.

Instance There is a relationship between the angle of deviation δ (the angle between the input beam and the output beam) and the refractive index of the prism medium for each wavelength of light [24]. We limit

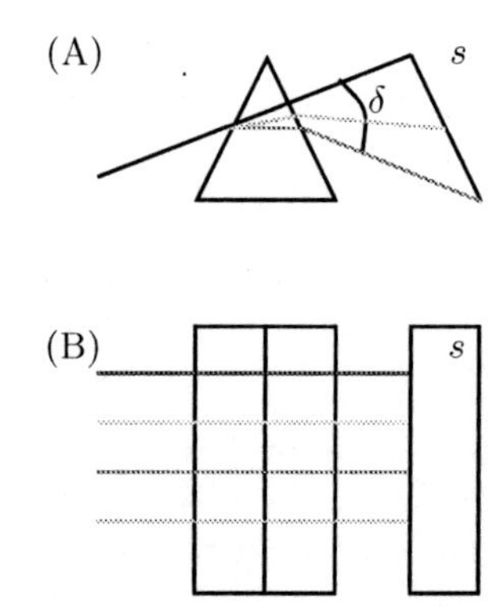

Fig. 5. Rainbow Sort and Generalised Rainbow Sort. (A) The computation of Rainbow Sort and also a side elevation of Generalised Rainbow Sort. Here s represents a sensor. The angle δ is the angle of deviation. (B) Top down view of Generalised Rainbow Sort.

the set of available input beam wavelengths so that the distance from where the uninterrupted beam would have reached the sensor to where the diffracted beam reaches it is linear in both $\tan\delta$ and the distance between the sensor and the prism. This is expressed as

$$s = p \tan \delta \tag{9}$$

where s is the distance along the sensor (see Fig. 5) and p is the distance between the sensor and the prism surface. Equation (9) satisfies Equation (3) if we let the Restricted Model triple be $S = (m, p, 0)$ where m is the minimum wavelength whose refracted path can be measured by the implementation and p is as in Equation (9). Rainbow Sort cannot is not stable as it returns a sorted list of wavelengths and does not output indices. Thus Rainbow Sort is an instance of the Restricted Model.

Generalised Rainbow Sort sorts the input in a manner that is similar to Rainbow Sort but instead returns a list of indices, and naturally deals with repeated elements in the input. The resulting Generalised Rainbow Sort is a stable sort. Using Equation (9) and the Model triple $S = (m, p, 0)$ from Section 3.4 we see that Generalised Rainbow Sort is an instance of the Model.

Implementation Here we show an implementation of Generalised Rainbow Sort. A list of numbers to be sorted $L =$ (635, 592, 513, 426, 513,

592, 426, 635, 426, 513, 592) was encoded as beams of light of a proportional wavelength. The beams were arranged in order, parallel to the axis face of a prism (see Fig. 6). The prism refracted the longer wavelengths to a lesser degree and so ordered them according to wavelength as shown in Fig. 7.

Reading off the beams in order of the most refracted and in the order of the list L yields the list of indices $c(L) = (4, 7, 9, 3, 5, 10, 2, 11, 1, 6, 8)$. This resulting list of numbers is a stable sorting of the list L.

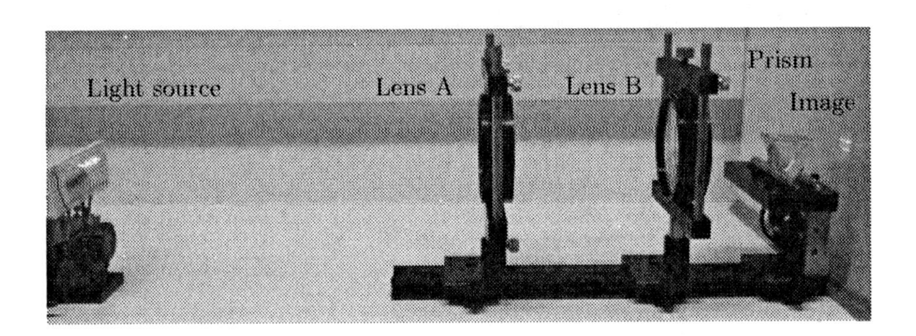

Fig. 6. The apparatus used to implement Generalised Rainbow Sort. Light emitting diodes were used as the light source. Lenses were used to focus the light beams onto the prism.

3.5 Mass spectrometry sort

Mass spectrometry [15] is a technique used for separating ions by their mass-to-charge ratio and is most commonly used to identify unknown compounds and to clarify the structure and chemical properties of molecules. Of the several types of mass spectrometry we describe here the "time of flight" method.

Description The process of time of flight mass spectrometric analysis is as follows. The gaseous sample particles are ionised by a short pulse of electrons and accelerated to a speed that is inversely proportional to their mass and directly proportional to their charge by a series of high voltage electric fields towards a long field-free vacuum tube known as the field-free drift region.

Here each ion moves at its entry velocity as they travel along the vacuum tube in a constant high voltage. At the opposite end of the tube

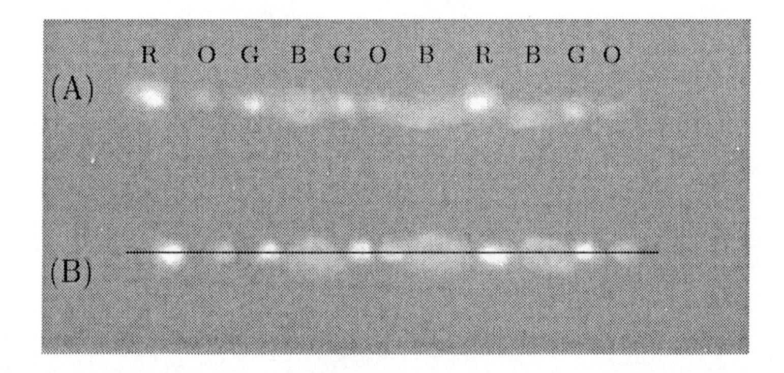

Fig. 7. (A) Beams that have been refracted proportional to their wavelength. Beams that are lower down than others have a lesser wavelength than those higher up. (B) Unrefracted light and the status of the beams before they were refracted by the prism. There is some distortion in the image introduced by the lenses. The coloured points are labeled G for green, O for orange, B for blue, and R for red.

there is a detector to record the arrival of the ions. In reflection time-of-flight mass spectrometers there is a "mirror" electric field which reflects the ions back along the length of the tube to the detector. This compensates for the initial energy spread of the ions and provides increased mass resolution. Since the different ions all travel the same distance but with characteristic velocities they arrive at the detector at different times. Using the time of arrival (time of flight) we identify the ions.

Sorting We refer to the use of mass spectrometry for sorting as Mass Spectrometry Sort. Given a list L of numbers to be sorted, we encode each element of L as a sample of molecules with a time of flight proportional to the element value. The samples of molecules are fired simultaneously by a mass spectrometer. The time of flight of each element is then recorded as it arrives at the sensor. This is the sorted list. Since we cannot record or distinguish multiple instances in the input Mass Spectrometry Sort is an unstable sort.

Our usual technique for introducing stability is to run identical instances that sort each element in parallel. However due to the cost of a mass spectrometer this is unfeasible.

Instance Gross [15] provides a simplified equation to describe the time of flight t of an ion based on its mass-to-charge ratio m_i/z is as follows

$$t = \frac{s}{\sqrt{2eU}}\sqrt{\frac{m_i}{z}} + t_0 \tag{10}$$

Where s is the distance traveled in the field-free region, U is the voltage of the field-free drift region, e is the charge of an electron and t_0 is a time offset. Equation (10) satisfies Equation (1) if we let $S = (m, \frac{s}{\sqrt{2eU}}, t_0)$ where m is the minimum time of flight that our apparatus measures. Our usual technique of running several apparatuses in parallel to achieve stability in the sorting is not practical due to expense. So in the way we have described it, Mass Spectrometry Sort is an instance of the Restricted Model of Physical sorting that is given in Definition 5.

Implementation As an example of Mass Spectrometry Sort we use the data from an existing mass spectrum [12] and interpret it as an instance of Mass Spectrometry Sort. If we are presented with an unordered list of integers $T = (19, 24, 9, 13, 19, 9, 14, 13)$ and we know the root mass-to-charge ratio for enough ions we map the values of each element of T to a root mean charge ratio. In this case we map the ion CH_3^+ to 9, CHO^+ to 13, CH_3O^+ to 14, CH_3OH^+ to 14, $(CH_3OH)_2^+$ to 19 and $(CH_3OH)_3^+$ to 24. The compounds necessary to create these ions are then placed in the spectrometric apparatus and ionised. The order that the compounds arrive at the detector is the sorted list and is shown in Fig. 8.

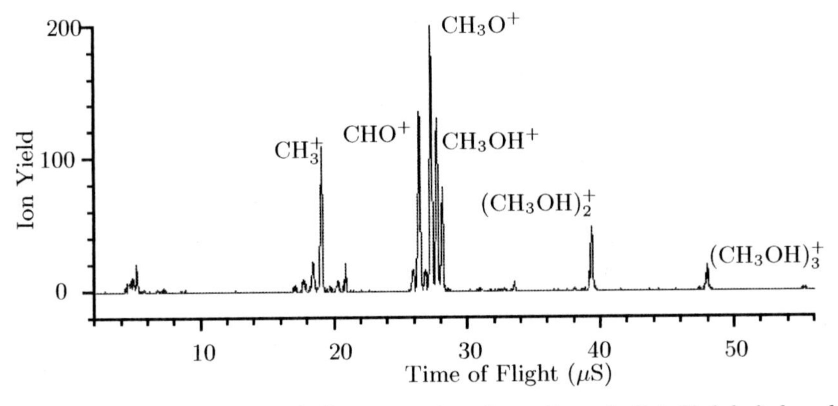

Fig. 8. A mass spectrum [12] representing the sorting of a list. Unlabeled peaks are not involved in the sort.

The time of arrival of each ion indicates the ion's identity and by reading the spectrograph the ordered list is found. In this case the sorted list is $c(T) = (19, 26, 27, 28, 39, 48)$ and since duplicates in the list have been lost it is clear that this is a non-stable sort.

4 Conclusion

In this paper we have proposed a Model of Physical Sorting that computes a stable sorting of its input list of natural numbers. This model has a parallel 1D to 2D (list to matrix) transformation as an atomic operation, where only one dimension of the matrix is dependent on the input list length. Once in matrix form it becomes a linear-time sequential task to read the list of stable sorted indices. We also define a Restricted Model of Physical Sorting which is an unstable sort.

We have provided five physical instances that are well-known laboratory techniques from experimental science as examples of physical sorts that are instances of the Model. We showed how the relationship between the Model and the Restricted Model naturally suggests how to introduce stability into a particular existing physics-inspired sort, that being Rainbow Sort. In four of the five instances we have presented experimental implementations. Several of the implementations has the potential to sort millions or more items in constant time. We assume however, the processes of encoding the input values and reading off the results are linear time bottlenecks. It might be possible that these bottlenecks be mitigated by parallelising these operations. Other candidate instances of the model that we have not considered here are centrifugal separation and fractional distillation [20].

There are many other sorting techniques available in the literature. All sequential comparison-based sorts have a lower bound of $\Omega(n \log n)$ comparisons [17] and some sequential non-comparison based sorting algorithms such as Radix Sort [17] and Counting Sort [17] have a worst case of $O(n \log n)$ time. Constant time parallel sorting is possible, however an unfeasible number of processors is required [13]. An approach from optical computing provides a constant time sort using a combination of lenses and individual sensors with on-board processing [18]. Other nature inspired sorting techniques such as Bead Sort [3] and Spaghetti Sort [2] are related to our Model of Physical Sorting and also have a linear time bound on sequentially reading the result. Several of our suggested implementations have the advantage that their technologies are already currently being used in research laboratories worldwide.

The practical advantages of our Model of Physical Sorting over traditional sorting algorithms would certainly not be apparent until the amount of objects to be sorted is in the billions. Even then, the implementation prospects of the Model could be in doubt. However, there are two reasons why we think the possibility should not be discounted completely. Firstly, several of the laboratory techniques underlying the implementations in this paper are widely used for other scientific purposes and significant resources are invested annually to improving the accuracy and reliability of the technologies. Each of these advances would benefit a sorting apparatus implemented using the same technology. Secondly, where an individual wishes to sort a set of particles by some property, it might be more efficient to directly use one of the technologies described in this paper rather than individually sensing the appropriate property of each particle, transferring it to a computer, and performing a traditional sort. The practicality of the Model of Physical sorting would also be less in doubt if there were implementations with a small footprint (for example 30cm^2) and were easily reusable.

5 Acknowledgments

This work was partially funded by the Irish Research Council for Science, Engineering and Technology. We would like to thank Charles Markham, William Lanigan, and Fiachra Matthews for their help in implementing Generalised Rainbow Sort. We would also like to thank reviewers for their helpful comments.

References

1. Adleman L., Molecular computation of solutions to combinatorial problems. *Science*, 266:1021–1024, 1994.
2. Dewdney A.K., On the spaghetti computer and other analog gadgets for problem solving. *Scientific American*, 250(6):19–26, June 1984.
3. Joshua J. Arulanandham, Cristian S. Calude, and Michael J. Dinneen., Bead-sort: A natural sorting algorithm. *The Bulletin of the European Association for Theoretical Computer Science*, 76:153–162, 2002.
4. Ashkin A., Acceleration and trapping of particles by radiation pressure. *Physical Review Letters*, 24(4):156–159, 1970.
5. Ashkin A., History of optical trapping and manipulation of small-neutral particle, atoms, and molecules. *IEEE Journal on Selected Topics in Quantum Electronics*, 6(6):841–856, 2000.
6. Ashkin A.,, J. M. Dziedzic, J. E. Bjorkholm, and Steven Chu, Observation of a single-beam gradient force optical trap for dielectric particles. *Optics Letters*, 11(5):288–290, 1986.
7. Benenson Y., T. Paz-Elizur, R. Adar, E. Keinan, Z. Livneh, and E. Shapiro, Programmable and autonomous computing machine made of biomolecules. *Nature*, 414:430–434, 2001.

8. Calude C.S. and Păun G., *Computing with Cells and Atoms.* Taylor & Francis Publishers, London, 2001.
9. Chiou P.Y., A.T. Ohta, and M.C. Wu., Massively parallel manipulation of single cells and microparticles using optical images. *Nature*, 436:370–372, July 2005.
10. Isaac L. Chuang, Neil Gershenfeld, and Marc Kubinec., Experimental implementation of fast quantum searching. *Physical Review Letters*, 18(15):3408–3411, 1998.
11. De Lacy Costello B. and Adamatzky A., Experimental implementation of collision-based gates in Belousov-Zhabotinsky medium. *Chaos, Solitons and Fractals*, 25(3):535–544, 2005.
12. Duffy E.M., Studying a supersonic beam of clusters using a mass spectrometer. Master's thesis, National University of Ireland, Maynooth, jul 2005.
13. Gasarch W., Golub E., and Kruskal C., A survey of constant time parallel sorting. *Bulletin of the European Association for Theoretical Computer Science*, 72:84–103, 2000.
14. Glückstad J., Microfluidics: Sorting particles with light. *Nature Materials*, 3:9–10, 2004.
15. Gross J.H., *Mass Spectrometry: A Textbook*. Springer, 2004.
16. Head T., Formal language theory and DNA: an analysis of the generative capacity of specific recombinant behaviors. *Bulletin of Mathematical Biology*, 47(6):737–759, 1987.
17. Knuth D.E., *The Art of Computing Programming: Sorting and Searching*, volume 3. Addison-Wesley, second edition, 1998.
18. Louri A., Hatch, J.A. Jr., and Na J. A constant-time parallel sorting algorithm and its optical implementation using smart pixels. *Applied Optics*, 34(17):3087–3097, 1995.
19. Mead C., *Analog VLSI and Neural Systems.* Addison-Wesley, Reading, Massachusetts, 1989.
20. Meloan C.E., *Chemical Separations: principles, techniques, and experiments.* Wiley-Interscience, 1999.
21. Naughton T., Javadpour Z., Keating J., Klíma M., and Rott J., General-purpose acousto-optic connectionist processor. *Optical Engineering*, 38(7):1170–1177, 1999.
22. Poole C.F. and Schuette S.A., *Contemporary practice of chromatography.* Elsevier, 1984.
23. Păun G., Computing with membranes. *Journal of Computer and System Sciences*, 61(1):108–143, 2000.
24. Schultes D., Rainbow sort: Sorting at the speed of light. *Natural Computing*, 5(1):67–82, 2006.
25. Colin F.S. and Whittaker M., Editors, *Electrophoretic techniques.* Academic Press, London, 1983.
26. Toffoli T. What are nature's 'natural' ways of computing? In *PhysComp '92 – Proceedings of the Workshop on Physics of Computation*, pages 5–9, 1992.

27. Viney C. and Fenton R.A., Physics and gel electrophoresis: using terminal velocity to characterize molecular weight. *European Journal of Physics*, 19(6):575–580, 1998.
28. Woods D. and Naughton T.J. An optical model of computation. *Theoretical Computer Science*, 334(1–3):227–258, April 2005.

Conventional or unconventional: Is any computer universal? *Rigueurs et revers du calcul universel*[*]

Selim G. Akl

School of Computing, Queen's University
Kingston, Ontario, Canada K7L 3N6
akl@cs.queensu.ca,
http://www.cs.queensu.ca/home/akl

Abstract. An evolving computation is one whose characteristics vary during its execution. These variations have many different origins and can manifest themselves in several ways. Thus, for example, the parameters of a computation, such as the data it uses, may vary with time independently of the computational environment in which the computation is carried out. Alternatively, it may be that the data interact with one another during the computation thus changing each other's value irreversibly. In this paper we describe a number of evolving computational paradigms, such as computations with time-varying variables, interacting variables, time-varying complexity, and so on. We show that evolving computations demonstrate the impossibility of achieving universality in computing, be it conventional or unconventional.

1 Introduction

Il pensait que la cause universelle, ordinatrice et première était bonne.
Denis Diderot

The universe in which we live is in a constant state of evolution. People age, trees grow, the weather varies. From one moment to the next, our world undergoes a myriad of transformations. Many of these changes

* This research was supported by the Natural Sciences and Engineering Research Council of Canada.

are obvious to the naked eye, others more subtle. Deceptively, some appear to occur independently of any direct external influences. Others are immediately perceived as the result of actions by other entities.

In the realm of computing, it is generally assumed that the world is static. The vast majority of computations take place in applications where change is thought of, rightly or wrongly, as inexistent or irrelevant. Input data are read, algorithms are applied to them, and results are produced. The possibility that the data, the algorithms, or even the results sought may vary *during* the process of computation is rarely, if ever, contemplated.

In this paper we explore the concept of *evolving computational systems*. These are systems in which everything in the computational process is subject to change. This includes inputs, algorithms, outputs, and even the computing agents themselves. A simple example of a computational paradigm that meets this definition of an evolving system to a limited extent is that of a computer interacting in real time with a user while processing information. Our focus here is primarily on certain changes that may affect the data required to solve a problem. We also examine changes that affect the complexity of the algorithm used in the solution. Finally, we look at one example of a computer capable of evolving with the computation.

A number of evolving computational paradigms are described. In Sections 3, 4, and 5, time plays an important role either directly or indirectly in the evolution of the computation. Thus, it is the passage of time that may cause the change in the data. In another context, it may be the order in which a stage of an algorithm is performed, that determines the number of operations required by that stage. In Sections 6 and 7, it is not time but rather external agents acting on the data that are responsible for a variable computation. Thus, the data may be affected by a measurement that perturbs an existing equilibrium, or by a modification in a mathematical structure that violates a required condition. Finally, in Section 8 evolving computations allow us to demonstrate that no computer, whether conventional or unconventional, can aspire to the title of 'universal', so long as its properties are fixed once and for all. Our conclusions are offered in Section 9.

2 Computational models

Like water runs around the rounding stone
time swims around the smoothing self
that polished becomes nothing but shine.
Gary Lark

It is appropriate at the outset that we define our models of computation. Two generic models are introduced in this section, one conventional and one unconventional. (A third model, a particular unconventional computer – the accelerating machine – is defined in Section 4.3.) We begin by stating clearly our understanding regarding the meaning of time, and our assumptions in connection with the speed of processors.

2.1 Time and speed

Ô temps, suspends ton vol!
Alphonse de Lamartine

In the classical study of algorithms, the notion of a *time unit* is fundamental to the analysis of an algorithm's running time. A time unit is the smallest discrete measure of time. In other words, time is divided into consecutive time units that are indivisible. All events occur at the beginning of a time unit. Such events include, for example, a variable changing its value, a processor undertaking the execution of a step in its computation, and so on.

It is worth emphasizing that the length of a time unit is not an absolute quantity. Instead, the duration of a time unit is specified in terms of a number of factors. These include the parameters of the computation at hand, such as the rate at which the data are received, or the rate at which the results are to be returned. Alternatively, a time unit may be defined in terms of the speed of the processors available (namely, the single processor on a sequential computer and each processor on a parallel computer). In the latter case, a faster processor implies a smaller time unit.

In what follows the standard definition of time unit is adopted, namely: A time unit is the length of time required by a processor to perform a *step* of its computation. Specifically, during a time unit, a processor executes a step consisting of:

1. A *read* operation in which it receives a constant number of fixed-size data as input,

2. A *calculate* operation in which it performs a fixed number of constant-time *arithmetic* and *logical* calculations (such as adding two numbers, comparing two numbers, and so on), and
3. A *write* operation in which it returns a constant number of fixed-size data as output.

All other occurrences external to the processor (such as the data arrival rate, for example) will be set and measured in these terms. Henceforth, the term *elementary operation* is used to refer to a read, a calculate, or a write operation.

2.2 What does it mean to compute?

The history of the universe is, in effect,
a huge and ongoing quantum computation.
The universe is a quantum computer.
Seth Lloyd

An important characteristic of the treatment in this paper, is the broad perspective taken to define what it means *to compute.* Specifically, *computation* is a process whereby information is manipulated by, for example, acquiring it (input), transforming it (calculation), and transferring it (output). Any form of information processing (whether occurring spontaneously in nature, or performed on a computer built by humans) is a computation. Instances of computational processes include:

1. Measuring a physical quantity,
2. Performing an arithmetic or logical operation on a pair of numbers, and
3. Setting the value of a physical quantity,

to cite but a few. These computational processes themselves may be carried out by a variety of means, including, of course, conventional (electronic) computers, but also through physical phenomena [35], chemical reactions [1], and transformations in living biological tissue [42]. By extension, *parallel computation* is defined as the execution of several such processes of the same type simultaneously.

2.3 Conventional model

Our generic conventional model of computation is the *sequential computer*, commonly used in the design and analysis of sequential (also known as *serial*) algorithms. It consists of a single processor made up of circuitry for executing arithmetic and logical operations and a number of registers that serve as internal memory for storing programs and data. For our purposes, the processor is also equipped with an input unit and an output unit that allow it to receive data from, and send data to, the outside world, respectively.

During each time unit of a computation the processor can perform:

1. A read operation, that is, receive a constant number of fixed-size data as input,
2. A calculate operation, that is, execute a fixed number of constant-time calculations on its input, and
3. A write operation, that is, return a constant number of fixed-size results as output.

It is important to note here, that the read and write operations can be, respectively, from and to the model's internal memory. In addition, both the reading and writing may be, on occasion, from and to an external medium in the environment in which the computation takes place. Several incarnations of this model exist, in theory and in practice [40]. The result of this paper, to the effect that no finite computer is universal, applies to all variants.

2.4 Unconventional model

In order to capture the essence of unconventional computation, we take a *parallel computer* as its generic model. Our choice is quite appropriate as this model is representative of the widest possible range of contemplated unconventional computers. Indeed, parallelism in one way or another is at the very heart of most unconventional computers proposed to date, including for example, quantum computers, biological computers (in vivo and in vitro), analog neural networks, chemical computers, and so on. Furthermore, the computational problems to be studied in this paper, *require* a certain degree of parallelism for their successful completion. Therefore, only an unconventional computer capable of parallel computing has any hope of tackling these tasks. It is important to note, however, that our choice of a parallel computer as a generic unconventional model is only for illustration purposes: Our result regarding the impossibility of achieving universal computation applies independently of this choice. In this context, a parallel computer allows us to underscore the fact that each computational problem we present is indeed solvable, though not by any putative universal computer (whether conventional or unconventional). For each problem, a computer capable of n operations in a given time unit, and purporting to be universal, can perform a computation that demands this many operations in that time unit, *but not one requiring $n + 1$ or more operations.*

The parallel computer consists of n processors, numbered 1 to n, where $n \geq 2$. Each processor is of the type described in Section 2.3. The processors are connected in some fashion and are able to communicate with one another for exchanging data and results [2]. The exact nature of the communication medium among the processors is of no consequence to the results described in this paper.

During each time unit of a computation a processor can perform:

1. A read operation, that is, receive as input a constant number of fixed-size data,
2. A calculate operation, that is, execute a fixed number of constant-time calculations on its input, and
3. A write operation, that is, return as output a constant number of fixed-size results.

As with the sequential processor, the input can be received from, and the output returned to, either the internal memory of the processor or the outside world. In addition, a processor in a parallel computer may receive its input from, and return its output to, another processor.

2.5 A fundamental assumption

The analyses in this paper assume that all models of computation use the fastest processors possible (within the bounds established by theoretical physics). Specifically, no sequential computer exists that is faster than the one of Section 2.3, and similarly no parallel computer exists whose processors are faster than those of Section 2.4. Furthermore, no processor on the parallel computer of Section 2.4 is faster than the processor of the sequential computer of Section 2.3. This is the *fundamental assumption in parallel computation.* It is also customary to suppose that the sequential and parallel computers use identical processors. We adopt this convention throughout this paper, with a single exception: In Section 4.3 we assume that the unconventional computer is in fact capable of increasing its speed at every step (at a pre-established rate, so that the number of operations executable at every consecutive step is known a priori and fixed once and for all).

3 Time-varying variables

Le temps m'échappe et fuit;
Alphonse de Lamartine

For a positive integer n larger than 1, we are given n functions, each of one variable, namely, F_0, F_1, ..., F_{n-1}, operating on the n physical variables x_0, x_1, ..., x_{n-1}, respectively. Specifically, it is required to compute $F_i(x_i)$, for $i = 0, 1, \ldots, n-1$. For example, $F_i(x_i)$ may be equal to x_i^2.

What is unconventional about this computation, is the fact that the x_i are themselves functions that vary with time. It is therefore appropriate to write the n variables as

$$x_0(t), x_1(t), \ldots, x_{n-1}(t),$$

that is, as functions of the time variable t. It is important to note here that, while it is known that the x_i change with time, the actual functions that effect these changes are not known (for example, x_i may be a true random variable).

All the physical variables exist in their natural environment within which the computation is to take place. They are all available to be operated on at the beginning of the computation. Thus, for each variable $x_i(t)$, it is possible to compute $F_i(x_i(t))$, provided that a computer is available to perform the calculation (and subsequently return the result).

Recall that time is divided into intervals, each of duration one time unit. It takes one time unit to evaluate $F_i(x_i(t))$. The problem calls for computing $F_i(x_i(t))$, $0 \leq i \leq n-1$, at time $t = t_0$. In other words, once all the variables have assumed their respective values at time $t = t_0$, the functions F_i are to be evaluated for all values of i. Specifically,

$$F_0(x_0(t_0)), F_1(x_1(t_0)), \ldots, F_{n-1}(x_{n-1}(t_0)),$$

are to be computed. The fact that $x_i(t)$ changes with the passage of time should be emphasized here. Thus, if $x_i(t)$ is not operated on at time $t = t_0$, then after one time unit $x_i(t_0)$ becomes $x_i(t_0 + 1)$, and after two time units it is $x_i(t_0 + 2)$, and so on. Indeed, time exists as a fundamental fact of life. It is real, relentless, and unforgiving. Time cannot be stopped, much less reversed. (For good discussions of these issues, see [28,45].) Furthermore, for $k > 0$, not only is each value $x_i(t_0 + k)$ different from $x_i(t_0)$, but also the latter cannot be obtained from the former. We illustrate this behavior through an example from physics.

3.1 Quantum decoherence

L'homme n'a point de port, le temps n'a point de rive;
Il coule, et nous passons!
Alphonse de Lamartine

A binary variable is a mathematical quantity that takes exactly one of a total of two possible values at any given time. In the base 2 number system, these values are 0 and 1, and are known as *binary digits* or *bits*. Today's conventional computers use electronic devices for storing and manipulating bits. These devices are in either one or the other of two physical states at any given time (for example, two voltage levels), one representing 0, the other 1. We refer to such a device, as well as the digit it stores, as a *classical bit*.

In *quantum computing*, a bit (aptly called a quantum bit, or *qubit*) is both 0 and 1 at the same time. The qubit is said to be in a *superposition* of the two values. One way to implement a qubit is by encoding the 0 and 1 values using the spin of an electron (for example, clockwise, or "up" for 1, and counterclockwise, or "down" for 0). Formally, a qubit is a unit vector in a two-dimensional state space, for which a particular orthonormal basis, denoted by $\{|0\rangle, |1\rangle\}$ has been fixed. The two basis vectors $|0\rangle$ and $|1\rangle$ correspond to the possible values a classical bit can take. However, unlike classical bits, a qubit can also take many other values. In general, an arbitrary qubit can be written as a linear combination of the computational basis states, namely, $\alpha|0\rangle + \beta|1\rangle$, where α and β are complex numbers such that $|\alpha|^2 + |\beta|^2 = 1$.

Measuring the value of the qubit (that is, reading it) returns a 0 with probability $|\alpha|^2$ and a 1 with a probability $|\beta|^2$. Furthermore, the measurement causes the qubit to undergo *decoherence* (literally, to lose its coherence). When decoherence occurs, the superposition is said to collapse: any subsequent measurement returns the same value as the one obtained by the first measurement. The information previously held in the superposition is lost forever. Henceforth, the qubit no longer possesses its quantum properties and behaves as a classical bit [33].

There is a second way, beside measurement, for decoherence to take place. A qubit loses its coherence simply through prolonged exposure to its natural environment. The interaction between the qubit and its physical surroundings may be thought of as an external action by the latter causing the former to behave as a classical bit, that is, to lose all information it previously stored in a superposition. (One can also view decoherence as the act of the qubit making a mark on its environment by adopting a classical value.) Depending on the particular implementation of the qubit, the time needed for this form of decoherence to take place varies. At the time of this writing, it is well below one second (more precisely, in the vicinity of a nanosecond). The information lost through decoherence cannot be retrieved. For the purposes of this example, the time required for decoherence to occur is taken as one time unit.

Now suppose that a quantum system consists of n independent qubits, each in a state of superposition. Their respective values at some time t_0, namely, $x_0(t_0)$, $x_1(t_0)$, ..., $x_{n-1}(t_0)$, are to be used as inputs to the n functions F_0, F_1, ..., F_{n-1}, in order to perform the computation described at the beginning of Section 3, that is, to evaluate $F_i(x_i(t_0))$, for $0 \leq i \leq n-1$.

3.2 Conventional solution

Le bonheur, c'est quand le temps s'arrête.
Gilbert Cesbron

A sequential computer fails to compute all the F_i as desired. Indeed, suppose that $x_0(t_0)$ is initially operated upon. It follows that $F_0(x_0(t_0))$ can be computed correctly. However, when the next variable, x_1, for example, is to be used (as input to F_1), the time variable would have changed from $t = t_0$ to $t = t_0 + 1$, and we obtain $x_1(t_0+1)$, instead of the $x_1(t_0)$ that we need. Continuing in this fashion, $x_2(t_0 + 2)$, $x_3(t_0 + 3)$, ..., $x_{n-1}(t_0 + n - 1)$, represent the sequence of inputs. In the example of Section 3.1, by the time $F_0(x_0(t_0))$ is computed, one time unit would have passed. At this point, the $n - 1$ remaining qubits would have undergone decoherence. The same problem occurs if the sequential computer attempts to first read all the x_i, one by one, and store them before calculating the F_i.

Since the function according to which each x_i changes with time is not known, it is impossible to recover $x_i(t_0)$ from $x_i(t_0 + i)$, for $i = 1$, 2, ..., $n - 1$. Consequently, this approach cannot produce $F_1(x_1(t_0))$, $F_2(x_2(t_0))$, ..., $F_{n-1}(x_{n-1}(t_0))$, as required.

3.3 Unconventional solution

La montre molle est une invention de Salvador Dali,
particulièrement adaptée aux horaires souples et aux journées élastiques,
mais inutilisable quand les temps sont durs.
Marc Escayrol

For a given n, *any* computer capable of performing n calculate operations per step, can easily evaluate the $F_i(x_i(t_0))$, all simultaneously, leading to a successful computation.

Thus, a parallel computer consisting of n independent processors may perform all the computations at once: For $0 \leq i \leq n-1$, and all processors working at the same time, processor i computes $F_i(x_i(t_0))$. In the example of Section 3.1, the n functions are computed in parallel at time $t = t_0$, before decoherence occurs.

4 Time-varying computational complexity

We are time's subjects, and time bids be gone.
William Shakespeare

In traditional computational complexity theory, the *size* of a problem $\mathcal{P}$ plays an important role. If $\mathcal{P}$ has size n, for example, then the number of operations required in the worst case to solve $\mathcal{P}$ (by any algorithm) is expressed as a function of n. Similarly, the number of operations executed (in the best, average, and worst cases) by a specific algorithm that solves $\mathcal{P}$ is also expressed as a function of n. Thus, for example, the problem of sorting a sequence of n numbers requires $\Omega(n \log n)$ comparisons, and the sorting algorithm Quicksort performs $O(n^2)$ comparisons in the worst case.

In this section we depart from this model. Here, the size of the problem plays a secondary role. In fact, in most (though not necessarily all) cases, the problem size may be taken as constant. The computational complexity now depends on *time*. Not only science and technology, but also everyday life, provide many instances demonstrating time-varying complexity. Thus, for example:

1. An illness may get better or worse with time, making it more or less amenable to treatment.
2. Biological and software viruses spread with time making them more difficult to cope with.
3. Spam accumulates with time making it more challenging to identify the legitimate email "needles" in the "haystack" of junk messages.
4. Tracking moving objects becomes harder as they travel away from the observer (for example, a spaceship racing towards Mars).
5. Security measures grow with time in order to combat crime (for example, when protecting the privacy, integrity, and authenticity of data, ever stronger cryptographic algorithms are used, that is, ones that are more computationally demanding to break, thanks to their longer encryption and decryption keys).
6. Algorithms in many applications have complexities that vary with time from one time unit during the computation to the next. Of particular importance here are:

(a) *Molecular dynamics* (the study of the dynamic interactions among the atoms of a system, including the calculation of parameters such as forces, energies, and movements) [18, 39], and
(b) *Computational fluid dynamics* (the study of the structural and dynamic properties of moving objects, including the calculation of the velocity and pressure at various points) [11].

Suppose that we are given an algorithm for solving a certain computational problem. The algorithm consists of a number of stages, where each stage may represent, for example, the evaluation of a particular arithmetic expression (such as $c \leftarrow a + b$). Further, let us assume that a computational stage executed at time t requires a number $C(t)$ of constant-time operations. As the aforementioned situations show, the behavior of C varies from case to case. Typically, C may be an increasing, decreasing, unimodal, periodic, random, or chaotic function of t. In what follows we study the effect on computational complexity of a number of functions $C(t)$ that *grow* with time.

It is worth noting that we use the term *stage* to refer to a component of an algorithm, hence a variable entity, in order to avoid confusion with a *step*, an intrinsic property of the computer, as defined in Sections 2.1 and 4.3. In conventional computing, where computational complexity is invariant (that is, oblivious to external circumstances), a *stage* (as required by an algorithm) is exactly the same thing as a *step* (as executed by a computer). In *unconventional computing* (the subject of this paper), computational complexity is affected by its environment and is therefore variable. Under such conditions, one or more steps may be needed in order to execute a stage.

4.1 Examples of increasing functions $C(t)$

The fundamental things apply
As time goes by.
Herman Hupfeld

Consider the following three cases in which the number of operations required to execute a computational stage increases with time. For notational convenience, we use $S(i)$ to express the number of operations performed in executing stage i, at the time when that stage is in fact executed. Denoting the latter by t_i, it is clear that $S(i) = C(t_i)$.

1. For $t \geq 0$, $C(t) = t + 1$. Table 1 illustrates t_i, $C(t_i)$, and $S(i)$, for $1 \leq i \leq 6$. It is clear in this case that $S(i) = 2^{i-1}$, for $i \geq 1$. It follows that the total number of operations performed when executing all stages, from stage 1 up to and including stage i, is

Stage i	t_i	$C(t_i)$	$S(i)$
1	0	$C(0)$	1
2	0 + 1	$C(1)$	2
3	1 + 2	$C(3)$	4
4	3 + 4	$C(7)$	8
5	7 + 8	$C(15)$	16
6	15 + 16	$C(31)$	32
7	31 + 32	$C(63)$	64

Table 1. Number of operations required to complete stage i when $C(t) = t+1$.

$$\sum_{j=1}^{i} 2^{j-1} = 2^i - 1.$$

It is interesting to note that, while $C(t)$ is a linear function of the time variable t, for its part $S(i)$ grows exponentially with $i - 1$, where i is the number of stages executed so far. The effect of this behavior on the total number of operations performed is appreciated by considering the following example. When executing a computation requiring $\log n$ stages for a problem of size n, $2^{\log n} - 1 = n - 1$ operations are performed.

2. For $t \geq 0$, $C(t) = 2^t$. Table 2 illustrates t_i, $C(t_i)$, and $S(i)$, for $1 \leq i \leq 5$.

Stage i	t_i	$C(t_i)$	$S(i)$
1	0	$C(0)$	2^0
2	0 + 1	$C(1)$	2^1
3	1 + 2	$C(3)$	2^3
4	3 + 8	$C(11)$	2^{11}
5	11 + 2048	$C(2059)$	2^{2059}

Table 2. Number of operations required to complete stage i when $C(t) = 2^t$.

In this case, $S(1) = 1$, and for $i > 1$, we have:

$$S(i) = 2^{\sum_{j=1}^{i-1} S(j)}.$$

Since $S(i) > \sum_{j=1}^{i-1} S(j)$, the total number of operations required by i stages is less than $2S(i)$, that is, $O(S(i))$.
Here we observe again that while $C(t) = 2C(t-1)$, the number of operations required by $S(i)$, for $i > 2$, increases significantly faster than double those required by all previous stages combined.

3. For $t \geq 0$, $C(t) = 2^{2^t}$. Table 3 illustrates t_i, $C(t_i)$, and $S(i)$, for $1 \leq i \leq 3$.

Stage i	t_i	$C(t_i)$	$S(i)$
1	0	$C(0)$	2^{2^0}
2	0 + 2	$C(2)$	2^{2^2}
3	2 + 16	$C(18)$	$2^{2^{18}}$

Table 3. Number of operations required to complete stage i when $C(t) = 2^{2^t}$.

Here, $S(1) = 2$, and for $i > 1$, we have:

$$S(i) = 2^{2^{\sum_{j=1}^{i-1} S(j)}}.$$

Again, since $S(i) > \sum_{j=1}^{i-1} S(j)$, the total number of operations required by i stages is less than $2S(i)$, that is, $O(S(i))$.
In this example, the difference between the behavior of $C(t)$ and that of $S(i)$ is even more dramatic. Obviously, $C(t) = C(t-1)^2$, where $t \geq 1$ and $C(0) = 2$, and as such $C(t)$ is a fast growing function ($C(4) = 65536$, while $C(7)$ is represented with 39 decimal digits). Yet, $S(i)$ grows at a far more dizzying pace: Already $S(3)$ is equal to 2 raised to the power 4×65536.

The significance of these examples and their particular relevance in unconventional computation are illustrated by the paradigm in the following section.

4.2 Computing with deadlines

tiempo tiempo tiempo tiempo.
Era Era.
César Vallejo

Suppose that a certain computation requires that n functions, each of one variable, be computed. Specifically, let $f_0(x_0), f_1(x_1), \ldots, f_{n-1}(x_{n-1})$, be the functions to be computed. This computation has the following characteristics:

1. The n functions are entirely independent. There is no precedence whatsoever among them; they can be computed in any order.
2. Computing $f_i(x_i)$ at time t requires $C(t) = 2^t$ operations, for $0 \leq i \leq n-1$ and $t \geq 0$.
3. There is a deadline for reporting the results of the computations: All n values $f_0(x_0)$, $f_1(x_1)$, $\ldots$, $f_{n-1}(x_{n-1})$ must be returned by the end of the third time unit, that is, when $t = 3$.

It should be easy to verify that no sequential computer, capable of exactly one constant-time operation per step (that is, per time unit), can perform this computation for $n \geq 3$. Indeed, $f_0(x_0)$ takes $C(0) = 2^0 = 1$ time unit, $f_1(x_1)$ takes another $C(1) = 2^1 = 2$ time units, by which time three time units would have elapsed. At this point none of $f_2(x_2), f_3(x_3), \ldots, f_{n-1}(x_{n-1})$ would have been computed.

By contrast, an n-processor parallel computer solves the problem handily. With all processors operating simultaneously, processor i computes $f_i(x_i)$ at time $t = 0$, for $0 \leq i \leq n-1$. This consumes one time unit, and the deadline is met.

The example in this section is based on one of the three functions for $C(t)$ presented in Section 4.1. Similar analyses can be performed in the same manner for $C(t) = t + 1$ and $C(t) = 2^{2^t}$, as well as other functions describing time-varying computational complexity.

4.3 Accelerating machines

Avec le temps...
avec le temps, va, tout s'en va.
Léo Ferré

In order to put the result in Section 4.2 in perspective, we consider a variant on the models of computation described thus far. An *accelerating machine* is a computer capable of increasing the number of operations it can do at each successive step of a computation. This is an unconventional–though primarily theoretical–model with no existing implementation (to date!). It is widely studied in the literature on unconventional computing [10, 12, 14, 43, 44, 46]. The importance of the accelerating machine lies primarily in its role in questioning some long held beliefs regarding uncomputability [13] and universality [7].

It is important to note that the rate of acceleration is specified at the time the machine is put in service and remains the same for the lifetime of the machine. Thus, the number of operations that the machine can execute during the ith step, is known in advance and fixed permanently, for $i = 1, 2, \ldots$.

Suppose that an accelerating machine is available which can *double* the number of operations that it can perform at each step. Such a machine would be able to perform one operation in the first step, two operations in the second, four operations in the third, and so on. How would such an extraordinary machine fare with the computational problem of Section 4.2?

As it turns out, an accelerating machine capable of doubling its speed at each step, is unable to solve the problem for $n \geq 4$. It would compute $f_0(x_0)$, at time $t = 0$ in one time unit. Then it would compute $f_1(x_1)$, which now requires two operations at $t = 1$, also in one time unit. Finally, $f_2(x_2)$, requiring four operations at $t = 2$, is computed in one time unit, by which time $t = 3$. The deadline has been reached and none of $f_3(x_3), f_4(x_4), \ldots, f_{n-1}(x_{n-1})$ has been computed.

In closing this discussion of accelerating machines we note that once an accelerating machine has been defined, a problem can always be devised to expose its limitations. Thus, let the acceleration function be $\Phi(t)$. In other words, $\Phi(t)$ describes the number of operations that the accelerating machine can perform at time t. For example, $\Phi(t) = 2\Phi(t-1)$, with $t \geq 1$ and $\Phi(0) = 1$, as in the case of the accelerating machine in this section. By simply taking $C(t) > \Phi(t)$, the accelerating machine is rendered powerless, *even in the absence of deadlines.*

5 Rank-varying computational complexity

Dans l'ordre naturel comme dans l'ordre social,
il ne faut pas vouloir être plus qu'on ne peut.
Nicolas de Chamfort

Unlike the computations in Section 4, the computations with which we are concerned here have a complexity that does not vary with time. Instead, suppose that a computation consists of n stages. There may be a certain precedence among these stages, that is, the order in which the stages are performed matters since some stages may depend on the results produced by other stages. Alternatively, the n stages may be totally independent, in which case the order of execution is of no consequence to the correctness of the computation.

Let the *rank* of a stage be the order of execution of that stage. Thus, stage i is the ith stage to be executed. In this section we focus on computations with the property that the number of operations required to execute a stage whose rank is i is a function of i only. For example, as in Section 4, this function may be increasing, decreasing, unimodal, random, or chaotic. Instances of algorithms whose computational complexity varies from one stage to another are described in [15]. As we did before, we concentrate here on the case where the computational complexity C is an *increasing* function of i.

When does rank-varying computational complexity arise? Clearly, if the computational requirements grow with the rank, this type of complexity manifests itself in those circumstances where it is a disadvantage, whether avoidable or unavoidable, to being ith, for $i \geq 2$. For example:

1. A penalty may be charged for missing a deadline, such as when a stage s must be completed by a certain time d_s.
2. The precision and/or ease of measurement of variables involved in the computation in a stage s may decrease with each stage executed before s.
3. Biological tissues may have been altered (by previous stages) when stage s is reached.
4. The effect of $s - 1$ quantum operations may have to be reversed to perform stage s.

5.1 An algorithmic example: Binary search

La fausse modestie consiste à se mettre sur le même rang que les autres pour mieux montrer qu'on les dépasse.
Sully Prudhomme

Binary search is a well-known (sequential) algorithm in computer science. It searches for an element x in a sorted list L of n elements. In the worst case, binary search executes $O(\log n)$ stages. In what follows, we denote by $B(n)$ the total number of elementary operations performed by binary search (on a sequential computer), and hence its running time, in the worst case.

Conventionally, it is assumed that $C(i) = O(1)$, that is, each stage i requires the same constant number of operations when executed. Thus,

$B(n) = O(\log n)$. Let us now consider what happens to the computational complexity of binary search when we assume, unconventionally, that the computational complexity of every stage i increases with i. Table 4 shows how $B(n)$ grows for three different values of $C(i)$.

$C(i)$	$B(n)$
i	$O(\log^2 n)$
2^i	$O(n)$
2^{2^i}	$O(2^n)$

Table 4. Number of operations required by binary search for different functions $C(i)$.

In a parallel environment, where n processors are available, the fact that the sequence L is sorted is of no consequence to the search problem. Here, each processor reads x, compares one of the elements of L to x, and returns the result of the comparison. This requires one time unit. Thus, regardless of $C(i)$, the running time of the parallel approach is always the same.

5.2 The inverse quantum Fourier transform

Je ne comprends pas qu'on laisse entrer les spectateurs des six premiers rangs avec des instruments de musique.

Alfred Jarry

Consider a quantum register consisting of n qubits. There are 2^n computational basis vectors associated with such a register, namely,

$$\begin{aligned} |0\rangle &= |000\cdots 00\rangle, \\ |1\rangle &= |000\cdots 01\rangle, \\ &\vdots \\ |2^n - 1\rangle &= |111\cdots 11\rangle. \end{aligned}$$

Let $|j\rangle = |j_1 j_2 j_3 \cdots j_{n-1} j_n\rangle$, be one of these vectors. For $j = 0, 1, \ldots, 2^n - 1$, the quantum Fourier transform of $|j\rangle$ is given by

$$\frac{(|0\rangle + e^{2\pi i 0.j_n}|1\rangle) \otimes (|0\rangle + e^{2\pi i 0.j_{n-1}j_n}|1\rangle) \otimes \cdots \otimes (|0\rangle + e^{2\pi i 0.j_1 j_2 \cdots j_n}|1\rangle)}{2^{n/2}},$$

where

1. Each transformed qubit is a balanced superposition of $|0\rangle$ and $|1\rangle$,
2. For the remainder of this section $i = \sqrt{-1}$,
3. The quantities $0.j_n$, $0.j_{n-1}j_n$, ..., $0.j_1j_2\cdots j_n$, are binary fractions, whose effect on the $|1\rangle$ component is called a *rotation*, and
4. The operator $\otimes$ represents a tensor product; for example,

$$(a_1|0\rangle + b_1|1\rangle)\otimes(a_2|0\rangle + b_2|1\rangle) = a_1a_2|00\rangle + a_1b_2|01\rangle + b_1a_2|10\rangle + b_1b_2|11\rangle.$$

We now examine the inverse operation, namely, obtaining the original vector $|j\rangle$ from its given quantum Fourier transform.

Conventional solution Since the computation of each of $j_1, j_2, \ldots j_{n-1}$ depends on j_n, we must begin by computing the latter from $|0\rangle + e^{2\pi i 0.j_n}|1\rangle$. This takes one operation. Now j_n is used to compute j_{n-1} from $|0\rangle + e^{2\pi i 0.j_{n-1}j_n}|1\rangle$ in two operations. In general, once j_n is available, j_k requires knowledge of j_{k+1},, j_{k+2}, ..., j_n, must be computed in $(n-k+1)$st place, and costs $n-k+1$ operations to retrieve from $|0\rangle + e^{2\pi i 0.j_kj_{k+1}\cdots j_n}|1\rangle$, for $k = n-1, n-2, \ldots, 1$. Formally, the sequential algorithm is as follows:

for $k = n$ **downto** 1 **do**

$$|j_k\rangle \leftarrow \frac{1}{\sqrt{2}}\begin{pmatrix} |0\rangle \\ e^{2\pi i 0.j_kj_{k+1}\cdots j_n}|1\rangle \end{pmatrix}$$

  for $m = k+1$ **to** n **do**

    if $j_{n+k+1-m} = 1$ **then**

$$|j_k\rangle \leftarrow |j_k\rangle \begin{pmatrix} 1 & 0 \\ 0 & e^{-2\pi i/2^{n-m+2}} \end{pmatrix}$$

    end if

  end for

$$|j_k\rangle \leftarrow |j_k\rangle \frac{1}{\sqrt{2}}\begin{pmatrix} 1 & 1 \\ 1 & -1 \end{pmatrix}$$

end for. ■

Note that the inner **for** loop is not executed when $m > n$. It is clear from the above analysis that a sequential computer obtains $j_1, j_2, \ldots, j_n$ in $n(n+1)/2$ time units.

Unconventional solution By contrast, a parallel computer can do much better in two respects. Firstly, for $k = n, n-1, \ldots, 2$, once j_k is known, all operations involving j_k in the computation of j_1, j_2, ..., j_{k-1}, can be performed simultaneously, each being a rotation. The parallel algorithm is given below:

for $k = 1$ **to** n **do in parallel**

$$|j_k\rangle \leftarrow \frac{1}{\sqrt{2}} \begin{pmatrix} |0\rangle \\ e^{2\pi i 0.j_k j_{k+1} \cdots j_n} |1\rangle \end{pmatrix}$$

end for

$$|j_n\rangle \leftarrow |j_n\rangle \frac{1}{\sqrt{2}} \begin{pmatrix} 1 & 1 \\ 1 & -1 \end{pmatrix}$$

for $k = n - 1$ **downto** 1 **do**

if $j_{k+1} = 1$ **then**

for $m = 1$ **to** k **do in parallel**

$$|j_m\rangle \leftarrow |j_m\rangle \begin{pmatrix} 1 & 0 \\ 0 & e^{-2\pi i/2^{n-m+1}} \end{pmatrix}$$

end for

end if

$$|j_k\rangle \leftarrow |j_k\rangle \frac{1}{\sqrt{2}} \begin{pmatrix} 1 & 1 \\ 1 & -1 \end{pmatrix}$$

end for. ■

The total number of time units required to obtain $j_1, j_2, \ldots, j_n$ is now $2n - 1$.

Secondly, and more importantly, if decoherence takes place within δ time units, where $2n - 1 < \delta < n(n+1)/2$, the parallel computer succeeds in performing the computation, while the sequential computer fails [34].

6 Interacting variables

If we take quantum theory seriously as a picture of what's really going on,
each measurement does more than disturb:
it profoundly reshapes the very fabric of reality.
Nick Herbert

So far, in every one of the paradigms that we have examined, the unconventional nature of the computation was due either to the passage of time or to the order in which an algorithmic stage is performed. In this and the next section, we consider evolving computations that occur in computational environments where time and rank play no role whatsoever either in the outcome or the complexity of the computation. Rather, it is the interactions among mutually dependent variables, caused by an interfering agent (performing the computation) that is the origin of the evolution of the system under consideration.

The computational paradigm to be presented in this section does have one feature in common with those discussed in the previous sections, namely, the central place occupied by the physical environment in which the computation is carried out. Thus, in Section 3, for example, the passage of time (a physical phenomenon, to the best of our knowledge) was the reason for the variables acquiring new values at each successive time unit. However, the attitude of the physical environment in the present paradigm is a passive one: Nature will not interfere with the computation until it is disturbed.

Let $\mathcal{S}$ be a physical system, such as one studied by biologists (for example, a living organism), or one maintained by engineers (for example, a power generator). The system has n variables each of which is to be measured or set to a given value at regular intervals. One property of $\mathcal{S}$ is that measuring or setting one of its variables modifies the values of any number of the system variables unpredictably. We show in this section how, under these conditions, a parallel solution method succeeds in carrying out the required operations on the variables of $\mathcal{S}$, while a sequential method fails. Furthermore, it is principles governing such fields as physics, chemistry, and biology, that are responsible for causing the inevitable failure of any sequential method of solving the problem at hand, while at the same time allowing a parallel solution to succeed. A typical example of such principles is the uncertainty involved in measuring several related variables of a physical system. Another principle expresses the way in which the components of a system in equilibrium react when subjected to outside stress.

6.1 Disturbing the equilibrium

All biologic phenomena act to adjust: there are no biologic actions other than adjustments. Adjustment is another name for Equilibrium. Equilibrium is Universal, or that which has nothing external to derange it.

Charles Fort

A physical system $\mathcal{S}$ possesses the following characteristics:

1. For $n > 1$, the system possesses a set of n variables (or properties), namely, $x_0, x_1, \ldots, x_{n-1}$. Each of these variables is a physical quantity (such as, for example, temperature, volume, pressure, humidity, density, electric charge, and so on). These quantities can be measured or set independently, each at a given discrete location (or point) within $\mathcal{S}$. Henceforth, x_i, $0 \leq i \leq n-1$, is used to denote a variable as well as the discrete location at which this variable is measured or set.

2. The system is in a state of *equilibrium*, meaning that the values x_0, x_1, ..., x_{n-1} satisfy a certain global condition $\mathcal{G}(x_0, x_1, \ldots, x_{n-1})$.
3. At regular intervals, the state of the physical system is to be recorded and possibly modified. In other words, the values $x_0, x_1, \ldots, x_{n-1}$ are to be measured at a given moment in time where $\mathcal{G}(x_0, x_1, \ldots, x_{n-1})$ is satisfied. New values are then computed for $x_0, x_1, \ldots, x_{n-1}$, and the variables are set to these values. Each interval has a duration of $\mathcal{T}$ time units; that is, the state of the system is measured and possibly updated every $\mathcal{T}$ time units, where $\mathcal{T} > 1$.
4. If the values $x_0, x_1, \ldots, x_{n-1}$ are measured or set *one by one*, each separately and independently of the others, this disturbs the equilibrium of the system. Specifically, suppose, without loss of generality, that all the values are first measured, and later all are set, in the order of their indices, such that x_0 is first and x_{n-1} last in each of the two passes. Thus:

   (a) When x_i is measured, an arbitrary number of values x_j, $0 \leq j \leq n-1$, will change unpredictably shortly thereafter (within one time unit), such that $\mathcal{G}(x_0, x_1, \ldots, x_{n-1})$ is no longer satisfied. Most importantly, when $i < n-1$, the values of $x_{i+1}, x_{i+2}, \ldots, x_{n-1}$, none of which has yet been registered, may be altered irreparably.

   (b) Similarly, when x_i is set to a new value, an arbitrary number of values x_j, $0 \leq j \leq n-1$, will change unpredictably shortly thereafter (within one time unit), such that $\mathcal{G}(x_0, x_1, \ldots, x_{n-1})$ is no longer satisfied. Most importantly, when $i > 0$, the values of $x_0, x_1, \ldots, x_{i-1}$, all of which have already been set, may be altered irreparably.

This last property of $\mathcal{S}$, namely, the way in which the system reacts to a sequential measurement or setting of its variables, is reminiscent of a number of well-known phenomena that manifest themselves in many subfields of the physical and natural sciences and engineering [8]. Examples of these phenomena are grouped into two classes and presented in what follows.

Uncertainty in measurement The phenomena of interest here occur in systems where measuring one variable of a given system affects, interferes with, or even precludes the subsequent measurement of another variable of the system. It is important to emphasize that the kind of uncertainty of concern in this context is in no way due to any errors that may be introduced by an imprecise or not sufficiently accurate measuring apparatus.

1. In quantum mechanics, *Heisenberg's uncertainty principle* puts a limit on our ability to measure pairs of 'complementary' variables. Thus, the *position* and *momentum* of a subatomic particle, or the *energy* of a particle in a certain state and the *time* during which that state existed, cannot be defined at the same time to arbitrary accuracy [9]. In fact, one may interpret this principle as saying that once *one* of the two variables is measured (however accurately, but independently of the other), the act of measuring itself introduces a disturbance that affects the value of the *other* variable. For example, suppose that at a given moment in time t_0 the position p_0 of an electron is measured. Assume further that it is also desired to determine the electron's momentum m_0 at time t_0. When the momentum is measured, however, the value obtained is not m_0, as it would have been changed by the previous act of measuring p_0.
2. In digital signal processing the *uncertainty principle* is exhibited when conducting a Fourier analysis. Complete resolution of a signal is possible either in the time domain t or the frequency domain w, but not both simultaneously. This is due to the fact that the Fourier transform is computed using e^{iwt}: Since the product wt must remain constant, narrowing a function in one domain, causes it to be wider in the other [19, 41]. For example, a pure sinusoidal wave has no time resolution, as it possesses nonzero components over the infinitely long time axis. Its Fourier transform, on the other hand, has excellent frequency resolution: It is an impulse function with a single positive frequency component. By contrast, an impulse (or *delta*) function has only one value in the time domain, and hence excellent resolution. Its Fourier transform is the constant function with nonzero values for all frequencies and hence no resolution.

Other examples in this class include image processing, sampling theory, spectrum estimation, image coding, and filter design [49]. Each of the phenomena discussed typically involves *two* variables in equilibrium. Measuring one of the variables has an impact on the value of the other variable. The system S, however, involves *several* variables (two or more). In that sense, its properties, as listed at the beginning of this section, are extensions of these phenomena.

Reaction to stress Phenomena in this class arise in systems where modifying the value of a parameter causes a change in the value of another parameter. In response to stress from the outside, the system automatically reacts so as to relieve the stress. Newton's third law of motion ("For every action there is an equal and opposite reaction") is a good way to characterize these phenomena.

1. In chemistry, *Le Châtelier's principle* states that if a system at equilibrium is subjected to a stress, the system will shift to a new equilibrium in an attempt to reduce the stress. The term *stress* depends on the system under consideration. Typically, stress means a change in pressure, temperature, or concentration [36]. For example, consider a container holding gases in equilibrium. Decreasing (increasing) the volume of the container leads to the pressure inside the container increasing (decreasing); in response to this external stress the system favors the process that produces the least (most) molecules of gas. Similarly, when the temperature is increased (decreased), the system responds by favoring the process that uses up (produces) heat energy. Finally, if the concentration of a component on the left (right) side of the equilibrium is decreased (increased), the system's automatic response is to favor the reaction that increases (decreases) the concentration of components on the left (right) side.
2. In biology, the *homeostatic principle* is concerned with the behavior displayed by an organism to which stress has been applied [37, 48]. An automatic mechanism known as *homeostasis* counteracts external influences in order to maintain the equilibrium necessary for survival, at all levels of organization in living systems. Thus, at the molecular level, homeostasis regulates the amount of enzymes required in metabolism. At the cellular level, it controls the rate of division in cell populations. Finally, at the organismic level, it helps maintain steady levels of temperature, water, nutrients, energy, and oxygen. Examples of homeostatic mechanisms are the sensations of hunger and thirst. In humans, sweating and flushing are automatic responses to heating, while shivering and reducing blood circulation to the skin are automatic responses to chilling. Homeostasis is also seen as playing a role in maintaining population levels (animals and their prey), as well as steady state conditions in the Earth's environment.

Systems with similar behavior are also found in cybernetics, economics, and the social sciences [25]. Once again, each of the phenomena discussed typically involves *two* variables in equilibrium. Setting one of the variables has an impact on the value of the other variable. The system $\mathcal{S}$, however, involves *several* variables (two or more). In that sense, its properties, as listed at the beginning of this section, are extensions of these phenomena.

6.2 Solutions

Que de temp perdu à gagner du temps.
Paul Morand

Two approaches are now described for addressing the problem defined at the beginning of Section 6.1, namely, to measure the state of $\mathcal{S}$ while in equilibrium, thus disturbing the latter, then setting it to a new desired state.

Simplifying assumptions In order to perform a concrete analysis of the different solutions to the computational problem just outlined, we continue to assume in what follows that the time required to perform all three operations below (in the given order) is one time unit:

1. Measuring one variable x_i, $0 \leq i \leq n-1$,
2. Computing a new value for a variable x_i, $0 \leq i \leq n-1$, and
3. Setting one variable x_i, $0 \leq i \leq n-1$.

Furthermore, once the new values of the parameters $x_0, x_1, \ldots, x_{n-1}$ have been applied to $\mathcal{S}$, the system requires one additional time unit to reach a new state of equilibrium. It follows that the smallest $\mathcal{T}$ can be is two time units; we therefore assume that $\mathcal{T} = 2$.

A mathematical model We now present a mathematical model of the computation in Section 6.1. Recall that the physical system has the property that all variables are related to, and depend on, one another. Furthermore, measuring (or setting) one variable disturbs any number of the remaining variables unpredictably (meaning that we cannot tell which variables have changed value, and by how much). Typically, the system evolves until it reaches a state of equilibrium and, in the absence of external perturbations, it can remain in a stable state indefinitely.

Formally, the interdependence among the n variables can be modeled using n functions, $g_0, g_1, \ldots, g_{n-1}$, as follows:

$$
\begin{aligned}
x_0(t+1) &= g_0(x_0(t), x_1(t), \ldots, x_{n-1}(t)) \\
x_1(t+1) &= g_1(x_0(t), x_1(t), \ldots, x_{n-1}(t)) \\
&\vdots \\
x_{n-1}(t+1) &= g_{n-1}(x_0(t), x_1(t), \ldots, x_{n-1}(t)).
\end{aligned}
$$

These equations describe the evolution of the system from state $(x_0(t), x_1(t), \ldots, x_{n-1}(t))$ at time t to state $(x_0(t+1), x_1(t+1), \ldots, x_{n-1}(t+1))$, one time unit later. While each variable is written as a function of time, there is a crucial difference between the present situation and that in Section 3: When the system is in a state of equilibrium, its variables do not change over time. It is also important to emphasize that, in most cases, the dynamics of the system are very complex, so the mathematical descriptions of functions $g_0, g_1, \ldots, g_{n-1}$ are either not known to us or we only have rough approximations for them.

Assuming the system is in an equilibrium state, our task is to measure its variables (in order to compute new values for these variables and set the system to these new values). In other words, we need the values of $x_0(t_0)$, $x_1(t_0)$, $\ldots$, $x_{n-1}(t_0)$ at moment $t = t_0$, when the system is in a stable state.

We can obtain the value of $x_0(t_0)$, for instance, by measuring that variable at time t_0 (noting that the choice of x_0 here is arbitrary; the argument remains the same regardless of which of the n variables we choose to measure first). Although we can acquire the value of $x_0(t_0)$ easily in this way, the consequences for the entire system can be dramatic. Unfortunately, any measurement is an external perturbation for the system, and in the process, the variable subjected to measurement will be affected unpredictably.

Thus, the measurement operation will change the state of the system from $(x_0(t_0), x_1(t_0), \ldots, x_{n-1}(t_0))$ to $(x'_0(t_0), x_1(t_0), \ldots, x_{n-1}(t_0))$, where $x'_0(t_0)$ denotes the value of variable x_0 after measurement. Since the measurement process has a non-deterministic effect upon the variable being measured, we cannot estimate $x'_0(t_0)$ in any way. Note also that the transition from $(x_0(t_0), x_1(t_0), \ldots, x_{n-1}(t_0))$, that is, the state before measurement, to $(x'_0(t_0), x_1(t_0), \ldots, x_{n-1}(t_0))$, that is, the state after measurement, does not correspond to the normal evolution of the system according to its dynamics described by functions g_i, $0 \leq i \leq n-1$.

However, because the equilibrium state was perturbed by the measurement operation, the system will react with a series of state transformations, governed by equations defining the g_i. Thus, at each time unit after t_0, the parameters of the system will evolve either towards a new equilibrium state or perhaps fall into a chaotic behavior. In any case, at time $t_0 + 1$, all n variables have acquired new values, according to the functions g_i:

$$
\begin{aligned}
x_0(t_0+1) &= g_0(x'_0(t_0), x_1(t_0), \ldots, x_{n-1}(t_0)) \\
x_1(t_0+1) &= g_1(x'_0(t_0), x_1(t_0), \ldots, x_{n-1}(t_0)) \\
&\vdots \\
x_{n-1}(t_0+1) &= g_{n-1}(x'_0(t_0), x_1(t_0), \ldots, x_{n-1}(t_0)).
\end{aligned}
$$

Consequently, unless we are able to measure all n variables, in parallel, at time t_0, some of the values composing the equilibrium state

$$(x_0(t_0), x_1(t_0), \ldots, x_{n-1}(t_0))$$

will be lost without any possibility of recovery.

Conventional approach The sequential computer measures *one* of the values (x_0, for example) and by so doing it disturbs the equilibrium, thus losing all hope of recording the state of the system within the given time interval. Any value read afterwards will not satisfy $\mathcal{G}(x_0, x_1, \ldots, x_{n-1})$.

Similarly, the sequential approach cannot update the variables of $\mathcal{S}$ properly: Once x_0 has received its new value, setting x_1 disturbs x_0 unpredictably.

Unconventional approach A parallel computer with n processors, by contrast, will measure *all* the variables $x_0, x_1, \ldots, x_{n-1}$ simultaneously (one value per processor), and therefore obtain an accurate reading of the state of the system within the given time frame. Consequently,

1. A snapshot of the state of the system that satisfies $\mathcal{G}(x_0, x_1, \ldots, x_{n-1})$ has been obtained.
2. The new variables $x_0, x_1, \ldots, x_{n-1}$ can be computed in parallel (one value per processor).
3. These new values can also be applied to the system simultaneously (one value per processor).

Following the resetting of the variables $x_0, x_1, \ldots, x_{n-1}$, a new equilibrium is reached. The entire process concludes within T time units successfully.

6.3 Distinguishability in quantum computing

Ne laissez jamais le temps au temps. Il en profite.
Jean Amadou

We conclude our study of interacting variables with an example from quantum computation. In Section 3.1 we saw that a single qubit can be in a superposition of two states, namely $|0\rangle$ and $|1\rangle$. In the same way, it is possible to place an entire quantum register, made up of n qubits, in a superposition of two states. The important point here is that, unlike the case in Section 3.1, it is not the individual qubits that are in a superposition, but rather the entire register (viewed as a whole).

Thus, for example, the register of n qubits may be put into any one of the following 2^n states:

$$\frac{1}{\sqrt{2}}(|000\cdots 0\rangle \pm |111\cdots 1\rangle)$$
$$\frac{1}{\sqrt{2}}(|000\cdots 1\rangle \pm |111\cdots 0\rangle)$$
$$\vdots$$
$$\frac{1}{\sqrt{2}}(|011\cdots 1\rangle \pm |100\cdots 0\rangle).$$

These vectors form an orthonormal basis for the state space corresponding to the n-qubit system. In such superpositions, the n qubits forming the system are said to be *entangled*: Measuring any one of them causes the superposition to collapse into one of the two basis vectors contributing to the superposition. Any subsequent measurement of the remaining $n - 1$ qubits will agree with that basis vector to which the superposition collapsed. This implies that it is impossible through single measurement to distinguish among the 2^n possible states. Thus, for example, if after one qubit is read the superposition collapses to $|000\cdots 0\rangle$, we will have no way of telling which of the two superpositions, $\frac{1}{\sqrt{2}}(|000\cdots 0\rangle + |111\cdots 1\rangle)$, or $\frac{1}{\sqrt{2}}(|000\cdots 0\rangle - |111\cdots 1\rangle)$, existed in the register prior to the measurement.

The only chance to differentiate among these 2^n states using quantum measurement(s) is to observe the n qubits simultaneously, that is, perform a single joint measurement of the entire system. In the given context, *joint* is really just a synonym for *parallel*. Indeed, the device in charge of performing the joint measurement must posses the ability to "read" the information stored in each qubit, in parallel, in a perfectly synchronized manner. In this sense, at an abstract level, the measuring apparatus can be viewed as having n probes. With all probes operating in parallel, each probe can "peek" inside the state of one qubit, in a perfectly synchronous operation. The information gathered by the n

probes is seen by the measuring device as a single, indivisible chunk of data, which is then interpreted to give one of the 2^n entangled states as the measurement outcome.

It is perhaps worth emphasizing that if such a measurement cannot be applied then the desired distinguishability can no longer be achieved regardless of how many other measuring operations we are allowed to perform. In other words, even an infinite sequence of measurements touching at most $n-1$ qubits at the same time cannot equal a single joint measurement involving all n qubits. Furthermore, with respect to the particular distinguishability problem that we have to solve, a single joint measurement capable of observing $n-1$ qubits simultaneously offers no advantage whatsoever over a sequence of $n-1$ consecutive *single* qubit measurements [31, 32].

7 Computations obeying mathematical constraints

The more constraints one imposes, the more one frees one's self. And the arbitrariness of the constraint serves only to obtain precision of execution.
Igor Stravinsky

In this section we examine computational problems in which a certain mathematical condition must be satisfied throughout the computation. Such problems are quite common in many subareas of computer science, such as numerical analysis and optimization. Thus, the condition may be a local one; for example, a variable may not be allowed to take a value larger than a given bound. Alternatively, the condition may be global, as when the average of a set of variables must remain within a certain interval. Specifically, for $n > 1$, suppose that some function of the n variables, $x_0, x_1, \ldots, x_i, \ldots, x_{n-1}$, is to be computed. The requirement here is that the variables satisfy a stated condition at each step of the computation. In particular, if the effect of the computation is to change x_i to x'_i at some point, then the condition must remain true, whether it applies to x_i alone or to the entire set of variables, whatever the case may be. If the condition is not satisfied at a given moment of the computation, the latter is considered to have failed.

Our concern in what follows is with computations that fit the broad definition just presented, yet can only be performed successfully in parallel (and not sequentially). All n variables, $x_0, x_1, \ldots, x_i, \ldots, x_{n-1}$, are already stored in memory. However, modifying any *one* of the variables from x_i to x'_i, to the exclusion of the others, causes the required condition (whether local or global) to be violated, and hence the computation to fail.

7.1 Mathematical transformations

Il n'y a que le temps qui ne perde pas son temps.
Jules Renard

There exists a family of computational problems where, given a mathematical object satisfying a certain property, we are asked to transform this object into another which also satisfies the same property. Furthermore, the property is to be maintained throughout the transformation, and be satisfied by every intermediate object, if any. Three examples of such transformations are now described.

Geometric flips The object shown in Fig. 1(a) is called a *convex subdivision*, as each of its faces is a convex polygon. This convex subdivision is to be transformed into that in Fig. 1(b).

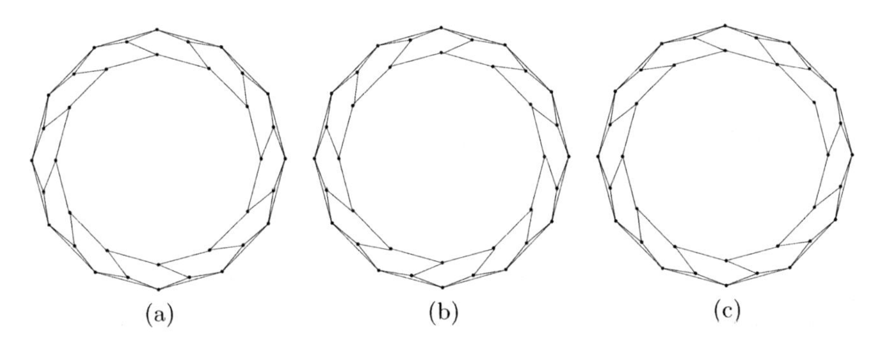

Fig. 1. Subdivision: (a) origin, (b) destination, (c) with a concavity.

The transformation can be effected by removing edges and replacing them with other edges. The condition for a successful transformation is that each intermediate figure (resulting from a replacement) be a convex subdivision as well. There are n edges in Fig. 1(a) that can be removed and replaced with another n edges to produce Fig. 1(b), where $n = 12$ for illustration. These are the "spokes" that connect the outside "wheel" to the inside one. However, as Fig. 1(c) illustrates, removing any *one* of these edges and replacing it with another creates a concavity, thus violating the condition [6, 29].

Map coloring A simple map is given consisting of n contiguous regions, where $n > 1$. Each region is a vertical strip going from the top edge to the bottom edge of the map. The regions are colored using two colors, red (R) and blue (B), in alternating fashion, thus:

$$\ldots RBRBRBRBRBRBRBRBRB \ldots$$

It is required to re-color this map, such that each region previously colored R is now colored B, and conversely, each region previously colored B is now colored R, thus:

$$\ldots BRBRBRBRBRBRBRBRBR \ldots$$

The condition to be satisfied throughout the recoloring is that no two adjacent regions are colored using the same color, and no third color (beside R and B) is ever used. It is clear that changing any *one* color at a time violates this requirement [24].

Rewriting systems From an initial string ab, in some formal language consisting of the two symbols a and b, it is required to generate the string $(ab)^n$, for $n > 1$. Thus, for $n = 3$, the target string is $ababab$. The rewrite rules to be used are:

$$a \rightarrow ab$$
$$b \rightarrow ab.$$

Throughout the computation, no intermediate string should have two adjacent identical characters. Such rewrite systems (also known as $\mathcal{L}$-systems) are used to draw fractals and model plant growth [38]. Here we note that applying any *one* of the two rules at a time causes the computation to fail (for example, if ab is changed to abb, by the first rewrite rule, or to aab by the second) [24].

7.2 Conventional solution

With all the x_i in its memory, suppose without loss of generality that the sequential computer obtains x'_0. This causes the computation to fail, as the set of variables x'_0, x_1, x_2, ..., x_{n-1} does not satisfy the global condition. Thus, in Section 7.1, only one edge of the subdivision in Fig. 1(a) can be replaced at a time. Once any one of the n candidate edges is replaced, the global condition of convexity no longer holds. The same is true in Sections 7.1 and 7.1, where the sequential computer can change only one color or one symbol at once, respectively, thereby causing the adjacency conditions to be violated.

7.3 Unconventional solution

For a given n, a parallel computer with n processors can easily perform a transformation on all the x_i collectively, with processor i computing x'_i. The required property in each case is maintained leading to a successful computation. Thus, in Section 7.1, n edges are removed from Fig. 1(a) and n new edges replace them to obtain Fig. 1(b), all in one step. Similarly in Section 7.1, all colors can be changed at the same time. Finally, in Section 7.1, the string $(ab)^n$ is obtained in $\log n$ steps, with the two rewrite rules being applied simultaneously to all symbols in the current intermediate string, in the following manner: *ab*, *abab*, *abababab*, and so on. It is interesting to observe that a successful generation of $(ab)^n$ also provides an example of a rank-varying computational complexity (as described in Section 5). Indeed, each legal string (that is, each string generated by the rules and obeying the adjacency property) is twice as long as its predecessor (and hence requires twice as many operations to be generated).

8 The universal computer is a myth

Every finitely realizable physical system can be perfectly simulated by a universal model computing machine operating by finite means.
David Deutsch

The Principle of Simulation is the cornerstone of computer science. It is at the heart of most theoretical results and practical implements of the field such as programming languages, operating systems, and so on. The principle states that any computation that can be performed on any one general-purpose computer can be equally carried out through simulation on any other general-purpose computer [17, 20, 30]. At times, the imitated computation, running on the second computer, may be faster or slower depending on the computers involved. In order to avoid having to refer to different computers when conducting theoretical analyses, it is a generally accepted approach to define a model of computation that can simulate all computations by other computers. This model would be known as a Universal Computer $\mathcal{U}$. Thus, Universal Computation, which clearly rests on the Principle of Simulation, is also one of the foundational concepts in the field [4, 16, 21].

Our purpose here is to prove the following general statement: There does not exist a *finite* computational device that can be called a Universal Computer. Our reasoning proceeds as follows. Suppose there exists a Universal Computer capable of n elementary operations per step, where

n is a finite and fixed integer. This computer will fail to perform a computation *requiring* n' operations per step, for any $n' > n$, and consequently lose its claim of universality. Naturally, for each $n' > n$, another computer capable of n' operations per step will succeed in performing the aforementioned computation. However, this new computer will in turn be defeated by a problem requiring $n'' > n'$ operations per step.

This reasoning is supported by each of the computational problems presented in Sections 3–7. As we have seen, these problems *can* easily be solved by a computer capable of executing n operations at every step. Specifically, an n-processor parallel computer led to a successful computation in each case. However, *none* of these problems is solvable by any computer capable of at most $n-1$ operations per step, for any integer $n > 1$. Furthermore, the problem size n itself is a variable that changes with each problem instance. As a result, *no* parallel computer, regardless of how many processors it has available, can cope with a growing problem size, as long as the number of processors is finite and fixed. This holds even if the computer purporting to be universal is endowed with an unlimited memory and is allowed to compute for an indefinite amount of time.

The preceding reasoning applies to any computer that obeys the *finiteness condition*, that is, a computer capable of only a finite and fixed number of operations per step. It should be noted that computers obeying the finiteness condition include all "reasonable" models of computation, both theoretical and practical, such as the Turing Machine [26], the Random Access Machine [40], and other idealized models, as well as all of today's general-purpose computers, including existing conventional computers (both sequential and parallel), as well as contemplated unconventional ones such as biological and quantum computers [5]. It is clear from Section 4.3 that even accelerating machines are not universal.

Therefore, the Universal Computer $\mathcal{U}$ is clearly a myth. As a consequence, the Principle of Simulation itself (though it applies to most *conventional* computations) is, in general, a fallacy. In fact, the latter principle is responsible for many other myths in the field. Of particular relevance to parallel computing, are the myths of the *Speedup Theorem* (speedup is at most equal to the number of processors used in parallel), the *Slowdown Theorem*, also known as *Brent's Theorem* (when q instead of p processors are used, $q < p$, the slowdown is at most p/q), and Amdahl's Law (maximum speedup is inversely proportional to the portion of the calculation that is sequential). Each of these myths can be dispelled using the same computations presented in this paper. Other computations for dispelling these and other myths are presented in [4].

9 Conclusion

First, you know, a new theory is attacked as absurd;
then it is admitted to be true, but obvious and insignificant;
finally it is seen to be so important that its adversaries claim that they themselves discovered it.
William James

An evolving computation is one whose characteristics vary during its execution. In this paper, we used evolving computations to identify a number of computational paradigms involving problems whose solution necessitates the use of a parallel computer. These include computations with variables whose values change with the passage of time, computations whose computational complexity varies as a function of time, computations in which the complexity of a stage of the computation depends on the order of execution of that stage, computations with variables that interact with one another and hence change each other's values through physical processes occurring in nature, and computations subject to global mathematical constraints that must be respected throughout the problem solving process. In each case, n computational steps must be performed simultaneously in order for the computation to succeed. A parallel computer with n processors can readily solve each of these problems. No sequential computer is capable of doing so. Interestingly, this demonstrates that one of the fundamental principles in computing, namely, that any computation by one computer can be simulated on another, is invalid. None of the parallel solutions described in this paper can be simulated on a sequential computer, regardless of how much time and memory are allowed.

Another consequence of our analysis is that the concept of universality in computing is unachievable. For every putative universal computer $\mathcal{U}_1$ capable of $V(t)$ operations at time unit t, it is always possible to define a computation $\mathcal{P}_1$ requiring $W(t)$ operations at time unit t to be completed successfully, where $W(t) > V(t)$, for all t. While $\mathcal{U}_1$ fails, another computer $\mathcal{U}_2$ capable of $W(t)$ operations at time unit t succeeds in performing $\mathcal{P}_1$ (only to be defeated, in turn, by a computation $\mathcal{P}_2$ requiring more than $W(t)$ operations at time unit t). Thus, no finite computer can be universal. That is to say, no machine, defined once and for all, can do all computations possible on other machines. This is true regardless of how $V(t)$ is defined, so long as it is fixed : It may be a constant (as with all of today's computers), or grow with t (as with accelerating machines). The only possible universal computer would be one that is capable of an *infinite* number of operations per step. As pointed out in [5] the Universe satisfies this condition. This observation agrees with recent thinking to the effect that the Universe is a computer [23, 27, 47, 50]. As stated in [17]: "[T]hink of all our knowledge-generating processes, our whole culture and civilization, and all the thought processes in the minds of every individual, and indeed the entire evolving biosphere as well, as being a gigantic *computation*. The whole thing is executing a self-motivated, self-generating computer program."

References

1. A. Adamatzky, B. DeLacy Costello, and T. Asai, *Reaction-Diffusion Computers*, Elsevier Science, 2005.
2. S.G. Akl, *Parallel Computation: Models And Methods*, Prentice Hall, 1997.
3. S.G. Akl, The design of efficient parallel algorithms, in: *Handbook on Parallel and Distributed Processing*, Blazewicz, J., Ecker, K., Plateau, B., and Trystram, D., Eds., Springer Verlag, 2000, pp. 13–91.
4. S.G. Akl, Superlinear performance in real-time parallel computation, *The Journal of Supercomputing*, Vol. 29, No. 1, 2004, pp. 89–111.
5. S.G. Akl, The myth of universal computation, in: *Parallel Numerics*, Trobec, R., Zinterhof, P., Vajteršic, M., and Uhl, A., Eds., Part 2, Systems and Simulation, University of Salzburg, Salzburg, Austria and Jožef Stefan Institute, Ljubljana, Slovenia, 2005, pp. 211–236.
6. S.G. Akl, Inherently parallel geometric computations, *Parallel Processing Letters*, Vol. 16, No. 1, March 2006, pp. 19–37.
7. S.G. Akl, Even accelerating machines are not universal, to appear in *International Journal of Unconventional Computing*.
8. S.G. Akl, B. Cordy, and W. Yao, An analysis of the effect of parallelism in the control of dynamical systems, *International Journal of Parallel, Emergent and Distributed Systems*, Vol. 20, No. 2, June 2005, pp. 147–168.
9. D.H. Bransden and C.J. Joachain, *Quantum Mechanics*, Pearson Education, 2000.
10. C.S. Calude and G. Păun, Bio-steps beyond Turing, *BioSystems*, Vol. 77, 2004, pp. 175–194.
11. T.J. Chung, *Computational Fluid Dynamics*, Cambridge University Press, 2002.
12. B.J. Copeland, Super Turing-machines, *Complexity*, Vol. 4, 1998, pp. 30–32.
13. B.J. Copeland, Even Turing machines can compute uncomputable functions, in: *Unconventional Models of Computation*, C.S. Calude, J. Casti, and M.J. Dinneen, Eds. Springer-Verlag, 1998, pp. 150–164.
14. B.J. Copeland, Accelerating Turing machines, *Mind and Machines*, Vol. 12, No. 2, 2002, pp. 281–301.
15. T.H. Cormen, C.E. Leiserson, R.L. Rivest, and C. Stein, *Introduction to Algorithms*, MIT Press, 2001.
16. M. Davis, *The Universal Computer*, W.W. Norton, 2000.

17. D. Deutsch, *The Fabric of Reality*, Penguin Books, 1997.
18. P.L. Freddolino, A.S. Arkhipov, S.B. Larson, A. McPherson, and K. Schulten, Molecular dynamics simulation of the complete satellite tobacco mosaic virus, *Structure*, Vol. 14, No. 3, March 2006, pp. pp. 437–449.
19. D. Gabor, Theory of communication, *Proceedings of the Institute of Electrical Engineers*, Vol. 93, No. 26, 1946, 420–441.
20. D. Harel, *Algorithmics: The Spirit of Computing*, Addison-Wesley, 1992.
21. D. Hillis, *The Pattern on the Stone*, Basic Books, 1998.
22. J.E. Hopcroft and J.D. Ullman, *Formal Languages and their Relations to Automata*, Addison-Wesley, 1969.
23. K. Kelly, God is the machine, *Wired*, Vol. 10, No. 12, December 2002.
24. A. Koves, personal communication, 2005.
25. R. Lewin, *Complexity*, The University of Chicago Press, 1999.
26. H.R. Lewis and C.H. Papadimitriou, *Elements of the Theory of Computation*, Prentice Hall, 1981.
27. S. Lloyd, *Programming the Universe*, Knopf, 2006.
28. M. Lockwood, *The Labyrinth of Time: Introducing the Universe*, Oxford University Press, 2005.
29. H. Meijer and D. Rappaport Simultaneous Edge Flips for Convex Subdivisions, *16th Canadian Conference on Computational Geometry*, Montreal, August 2004, pp. 57–69.
30. M.L. Minsky, *Computation: Finite and Infinite Machines*, Prentice-Hall, 1967.
31. M. Nagy and S.G. Akl, On the importance of parallelism for quantum computation and the concept of a universal computer, *Proceedings of the Fourth International Conference on Unconventional Computation*, Sevilla, Spain, October 2005, pp. 176–190.
32. M. Nagy and S.G. Akl, Quantum measurements and universal computation, *International Journal of Unconventional Computing*, Vol. 2, No. 1, 2006, pp. 73–88.
33. M. Nagy and S.G. Akl, Quantum computation and quantum information, *International Journal of Parallel, Emergent and Distributed Systems*, Vol. 21, No. 1, February 2006, pp. 1–59.
34. M. Nagy and S.G. Akl, Coping with decoherence: Parallelizing the quantum Fourier transform, Technical Report No. 2006-507, School of Computing, Queen's University, Kingston, Ontario, Canada, 2006.
35. M.A. Nielsen and I.L. Chuang, *Quantum Computation and Quantum Information*, Cambridge University Press, 2000.
36. L.W. Potts, *Quantitative Analysis*, Harper & Row, 1987.
37. D.B. Pribor, *Functional Homeostasis: Basis for Understanding Stress*, Kendall Hunt, 1986.
38. P. Prusinkiewicz and A. Lindenmayer, *The Algorithmic Beauty of Plants*, Springer Verlag, 1990.
39. D.C. Rapaport, *The Art of Molecular Dynamics Simulation*, Cambridge University Press, 2004.
40. J.E. Savage, *Models of Computation*, Addison-Wesley, 1998.

41. C.E. Shannon, Communication in the presence of noise, *Proceedings of the IRE*, Vol. 37, 1949, 10–21.
42. T. Sienko, A. Adamatzky, N.G. Rambidi, M. Conrad, Eds., *Molecular Computing*, MIT Press, 2003.
43. I. Stewart, Deciding the undecidable, *Nature*, Vol. 352, August 1991, pp. 664–665.
44. I. Stewart, The dynamics of impossible devices, *Nonlinear Science Today*, Vol. 1, 1991, pp. 8–9.
45. G. Stix, Real time, *Scientific American Special Edition: A Matter of Time*, Vol. 16, No. 1, February 2006, pp. 2–5.
46. K. Svozil, The Church-Turing thesis as a guiding principle for physics, in: *Unconventional Models of Computation*, C.S. Calude, J. Casti, and M.J. Dinneen, Eds. Springer-Verlag, 1998, pp. 150–164.
47. F.J. Tipler, *The Physics of Immortality*, Macmillan, 1995.
48. P.A. Trojan, *Ecosystem Homeostasis*, Dr. W. Junk, 1984.
49. R. Wilson and G.H. Granlund, The uncertainty principle in image processing, *IEEE Transactions on Pattern Analysis and Machine Intelligence*, Vol. PAMI-6, No. 6, 1984, 758–767.
50. S. Wolfram, *A New Kind of Science*, Wolfram Media, 2002.

Language diversity of measured quantum processes

Karoline Wiesner[1] and James P. Crutchfield[1]

Center for Computational Science & Engineering and Physics Department,
University of California Davis, One Shields Avenue, Davis, CA 95616 and
Santa Fe Institute, 1399 Hyde Park Road, Santa Fe, NM 87501
wiesner@cse.ucdavis.edu
chaos@cse.ucdavis.edu

Abstract. The behavior of a quantum system depends on how it is measured. How much of what is observed comes from the structure of the quantum system itself and how much from the observer's choice of measurement? We explore these questions by analyzing the *language diversity* of quantum finite-state generators. One result is a new way to distinguish quantum devices from their classical (stochastic) counterparts. While the diversity of languages generated by these two computational classes is the same in the case of periodic processes, quantum systems generally generate a wider range of languages than classical systems.

1 Introduction

Quantum computation has advanced dramatically from Feynman's initial theoretical proposal [1] to the experimental realizations one finds today. The largest quantum device that has been implemented, though, is a 7 qubit register that can factor a 3 bit number [2] using Shor's algorithm [3]. A review of this and other currently feasible quantum devices reveals that, for now and the foreseeable future, they will remain small—in the sense that a very limited number of qubits can be stored. Far from implementing the theoretical ideal of a quantum Turing machine, current experiments test quantum computation at the level of small finite-state machines.

The diversity of quantum computing devices that lie between the extremes of finite-state and (unbounded memory) Turing machines is substantially less well understood than, say, that for classical automata, as codified in the Chomsky hierarchy [4]. As an approach to filling in a quantum hierarchy, comparisons between classical and quantum automata can be quite instructive.

Such results are found for automata at the level of finite-state machines [5–7]. For example, the regular languages are recognized by finite-state machines (by definition), but quantum finite-state machines, as defined in Ref. [6], cannot recognize all regular languages. This does not mean, however, that quantum automata are strictly less powerful than their classical counterparts. There are nonregular languages that are recognized by quantum finite-state machines [8]. These first results serve to illustrate the need for more work, if we are to fully appreciate the properties of quantum devices even at the lowest level of some presumed future quantum computational hierarchy.

The comparison of quantum and classical automata has recently been extended to the probabilistic languages recognized by stochastic and quantum finite-state machines [7]. There, quantum finite-state generators were introduced as models of the behaviors produced by quantum systems and as tools with which to quantify their information storage and processing capacities.

Here we continue the effort to quantify information processing in simple quantum automata. We will show how a quantum system's possible behaviors can be characterized by the diversity of languages it generates under different measurement protocols. We also show how this can be adapted to measurements, suitably defined, for classical automata. It turns out that the diversity of languages, under varying measurement protocols, provides a useful way to explore how classical and quantum devices differ. A measured quantum system and its associated measured classical system can generate rather different sets of stochastic languages. For periodic processes, the language diversities are the same between the quantum and counterpart classical systems. However, for aperiodic processes quantum systems are more diverse, in this sense, and potentially more capable.

In the following, we first review formal language and automata theory, including stochastic languages, stochastic and quantum finite-state generators, and the connection between languages and behavior. We then introduce the *language diversity* of a finite-state automaton and analyze a number of example processes, comparing quantum and classical models. We conclude with a few summary remarks and contrast the language diversity with *transient information*, which measures the amount of information an observer needs to extract in order to predict which internal state a process is in [9].

2 Formal languages and behavior

Our use of formal language theory differs from most in how it analyzes the connection between a language and the systems that can generate it. In brief, we observe a system through a finite-resolution measuring instrument, representing each measurement with a *symbol* σ from discrete *alphabet* Σ. The temporal behavior of a system, then, is a string or a *word* consisting of a succession of measurement symbols. The collection of all (and only) those words is the *language* that captures the possible, temporal behaviors of the system.

Definition. *A* formal language $\mathcal{L}$ *is a set of* words $w = \sigma_0\sigma_1\sigma_2\ldots$ *each of which consists of a series of symbols* $\sigma_i \in \Sigma$ *from a discrete alphabet* Σ.

Σ^* denotes the set of all possible words of any length formed using symbols in Σ. We denote a word of length L by $\sigma^L = \sigma_0\sigma_1\ldots\sigma_{L-1}$, with $\sigma_i \in \Sigma$. The set of all words of length L is Σ^L.

Since a formal language, as we use the term, is a set of observed words generated by a process, then each *subword* $\sigma_i\sigma_{i+1}\cdots\sigma_{j-1}\sigma_j, i \leq j,\ i,j = 0,1,\ldots,L-1$ of a word σ^L has also been observed and is considered part of the language. This leads to the following definition.

Definition. *A language* $\mathcal{L}$ *is* subword closed *if, for each* $w \in \mathcal{L}$*, all of* w*'s subwords* $\mathrm{sub}(w)$ *are also members of* $\mathcal{L}$: $\mathrm{sub}(w) \subseteq \mathcal{L}$.

Beyond a formal language listing which words (or behaviors) occur and which do not, we are also interested in the probability of their occurrence. Let $\Pr(w)$ denote the probability of word w, then we have the following definition.

Definition. *A* stochastic language $\mathcal{L}$ *is a formal language with a* word distribution $\Pr(w)$ *that is normalized at each length* L:

$$\sum_{\{\sigma^L \in \mathcal{L}\}} \Pr(\sigma^L) = 1 \ , \tag{1}$$

with $0 \leq \Pr(\sigma^L) \leq 1$.

Definition. *Two stochastic languages* $\mathcal{L}_1$ *and* $\mathcal{L}_2$ *are said to be* δ*-similar if* $\forall \sigma^L \in \mathcal{L}_1$ *and* $\sigma'^L \in \mathcal{L}_2$: $|\Pr(\sigma^L) - \Pr(\sigma'^L)| \leq \delta$*, for all* L *and a specified* $0 \leq \delta \leq 1$*. If this is true for* $\delta = 0$*, then the languages are equivalent.*

For purposes of comparison between various computational models, it is helpful to refer directly to the set of words in a stochastic language $\mathcal{L}$. This is the *support* of a stochastic language:

$$\text{supp}(\mathcal{L}) = \{w \in \mathcal{L} : \ \Pr(w) > 0\} \ . \tag{2}$$

The support itself is a formal language. Whenever we compare formal and stochastic languages we add the respective subscripts and write $\mathcal{L}_{formal}$ and $\mathcal{L}_{stoch}$.

3 Stochastic finite-state generators

Automata with finite memory—*finite-state machines*—consist of a finite set of states and transitions between them [4]. Typically, they are used as *recognition* devices, whereas we are interested in the generation of words in a stochastic language. So here we will review models for classical and quantum generation, referring the reader to [10] for details on recognizers and automata in general.

Definition. *[7] A* stochastic generator *G is a tuple $\{S, Y, \{T(y)\}\}$ where*

1. *S is a finite set of states, with $|S|$ denoting its cardinality.*
2. *Y is a finite alphabet for output symbols.*
3. *$\{T(y), y \in Y\}$ is a set of $|Y|$ square stochastic matrices of order $|S|$. $|Y|$ is the cardinality of Y, the components $T_{ij}(y)$ give the probability of moving to state s_j and emitting y when in state s_i.*
4. *At each step a symbol $y \in Y$ is emitted and the machine updates its state. Thus, $\sum_{y \in Y} \sum_j T_{ij}(y) = 1$.*

Definition. *A* deterministic generator *(DG) is a G in which each matrix $T(y)$ has at most one nonzero entry per row.*

3.1 Process languages

Definition. *A* process language *$\mathcal{P}$ is a stochastic language that is subword closed.*

The output of a stochastic generator (as well as the quantum generator introduced below) is a process language; for the proof see Ref. [7]. Thus, all stochastic languages discussed in the following are process languages.

Definition. *A* periodic process language *with period N is a process language such that $\forall w = \sigma_0 \sigma_1 \ldots \sigma_n \in \mathcal{P}$ with $n \geq N$: $\sigma_i = \sigma_{i+N}$.*

Before discussing the languages associated with a G, we must introduce some helpful notation.

Notation. *Let* $|\eta\rangle = (11\ldots11)^T$ *denote a column vector with* $|S|$ *components that are all* 1*s.*

Notation. *The* state vector $\langle\pi| = (\pi_0, \pi_1, \ldots, \pi_{|S|-1})$ *is a row vector whose components,* $0 \leq \pi_i \leq 1$*, give the probability of being in state* s_i*. The state vector is normalized in probability:* $\sum_{i=0}^{|S|-1} \pi_i = 1$*. The* initial state distribution *is denoted* $\langle\pi^0|$.

The state-to-state transition probabilities of a G, independent of outputs, are given by the *state-to-state transition matrix*:

$$T = \sum_{y \in Y} T(y) , \tag{3}$$

which is a stochastic matrix: i.e., $0 \leq T_{ij} \leq 1$ and $\sum_j T_{ij} = 1$.

The generator updates its state distribution after each time step as follows:

$$\langle\pi^{t+1}| = \langle\pi^t|T(y) , \tag{4}$$

where (re)normalization of the state vector is assumed.

If a G starts in state distribution $\langle\pi^0|$, the probability of generating y^L is given by the state vector without renormalization

$$\Pr(y^L) = \langle\pi^0|T(y^L)|\eta\rangle , \tag{5}$$

where $T(y^L) = \prod_{i=0}^{L-1} T(y_i)$ represents the assumption in our model that all states are accepting. This, in turn, is a consequence of our focusing on process languages, which are subword closed.

4 Quantum generators

Quantum generators are a subset of *quantum machines* (or *transducers*), as defined in Ref. [7]. Their architecture consists of a set of internal states and transitions and an output alphabet that labels transitions. For simplicity here we focus on the definition of generators, without repeating the general definition of quantum transducers. Our basic *quantum generator* (*QG*) is defined as follows.

Definition. *[7] A QG is a tuple* $\{Q, \mathcal{H}, Y, \mathbf{T}(Y)\}$ *where*

1. $Q = \{q_i : i = 0, \ldots, n-1\}$ *is a set of* $n = |Q|$ internal states.
2. *The* state space $\mathcal{H}$ *is an* n*-dimensional Hilbert space.*

3. *The* state vector *is* $\langle\psi| \in \mathcal{H}$.
4. *Y is a finite alphabet for output symbols. $\lambda \notin Y$ denotes the null symbol.*
5. $\mathbf{T}(Y)$ *is a set of n-dimensional* transition matrices $\{T(y) = P(y) \cdot U, y \in Y\}$ *that are products of a unitary matrix U and a projection operator $P(y)$ where*
   (a) *U is an n-dimensional unitary* evolution operator *that governs the evolution of the state vector.*
   (b) $\mathbf{P}(Y)$ *is a set of n-dimensional* projection operators—$\mathbf{P} = \{P(y) : y \in Y \cup \{\lambda\}\}$*—that determines how a state vector is measured. The $P(y)$ are Hermitian matrices.*

At each time step a QG outputs a symbol $y \in Y$ or the null symbol λ and updates its state vector.

The output symbol y is identified with the measurement outcome. The symbol λ represents the event of no measurement. In the following we will concentrate on deterministic quantum generators. They are more transparent than general (nondeterministic) QGs, but still serve to illustrate the relative power of quantum and classical generators.

Definition. *A* quantum deterministic generator *(QDG) is a QG in which each matrix $T(y)$ has at most one nonzero entry per row.*

4.1 Observation and operation

The projection operators determine how output symbols are generated from the internal, hidden dynamics. In fact, the only way to observe a quantum process is to apply a projection operator to the current state. In contrast with classical processes, the measurement event disturbs the internal dynamics. The projection operators are familiar from quantum mechanics and can be defined in terms of the internal states as follows.

Definition. *A* projection operator $P(y)$ *is the linear operator*

$$P(y) = \sum_{\kappa \in \mathcal{H}_y} |\phi_\kappa\rangle\langle\phi_\kappa| \,, \qquad (6)$$

where κ runs over the indices of a one- or higher-dimensional subspace $\mathcal{H}_y$ of the Hilbert space and the ϕ_κ span these subspaces.

We can now describe a *QG*'s operation. U_{ij} is the transition amplitude from state q_i to state q_j. Starting in state $\langle\psi_0|$ the generator updates its state by applying the unitary matrix U. Then the state vector is projected using $P(y)$ and renormalized. Finally, symbol $y \in Y$ is emitted. In other words, a single time-step of a *QG* is given by:

$$\langle \psi(y)| = \langle \psi^0|UP(y) \ , \tag{7}$$

where (re)normalization of the state vector is assumed. The state vector after L time steps when emitting string y^L is

$$\langle \psi(y^L)| = \langle \psi^0| \prod_{i=0}^{L-1} (UP(y_i)) \ . \tag{8}$$

We can now calculate symbol and word probabilities of the process language generated by a *QG*. Starting the *QG* in $\langle \psi^0|$ the probability of output symbol y is given by the state vector without renormalization:

$$\Pr(y) = \|\psi(y)\|^2 \ . \tag{9}$$

By extension, the probability of output string y^L is

$$\Pr(y^L) = \left\|\psi(y^L)\right\|^2 \ . \tag{10}$$

4.2 Properties

In Ref. [7] we established a number of properties of *QG*s: their consistency with quantum mechanics, that they generate process languages, and their relation to stochastic generators and to quantum and stochastic recognizers. Here we avail ourselves of one property in particular of *QDG*s—for a given *QDG* there is always an equivalent (classical) deterministic generator. The latter is obtained by squaring the matrix elements of the *QDG*'s unitary matrix and using the same projection operators. The resulting state-to-state transition matrix is doubly stochastic; i.e., $0 \leq T_{ij} \leq 1$ and $\sum_i T_{ij} = \sum_j T_{ij} = 1$.

Theorem 1. *Every process language generated by a QDG is generated by some DG.*

Proof. *See Ref. [7].*

This suggests that the process languages generated by *QDG*s are a subset of those generated by *DG*s. In the following, we will take a slightly different perspective and ask what set of languages a given *QDG* can generate as one varies the *measurement protocol*—that is, the choice of measurements.

5 Language diversity

The notion of a measurement protocol is familiar from quantum mechanics: We define the *measurement period* as the number of applications of a projection operator relative to the unitary evolution time step. For a classical system this is less familiar, but it will be used in the same way. The measurement period here is the period of observing an output symbol relative to the internal state transitions. The internal dynamics remain unaltered in the classical case, whether the system is measured or not. In the quantum case, as is well known, the situation is quite different. Applying a projection operator disturbs the internal dynamics.

Definition. *A process observed with* measurement period p *is measured every* p *time steps.*

Note that this model of a measurement protocol, by which we subsample the output time series, is related to von Mises version of probability theory based on "collectives" [11].

The resulting observed behavior can be described in terms of the state-to-state transition matrix and the projection operators. For a classical finite-state machine this is:

$$\langle \pi(y)^{t+p}| = \langle \pi^t | T^{p-1} T(y) \ , \tag{11}$$

where $\langle \pi(y)^{t+p}|$ is the state distribution vector after p time steps and after observing symbol y. Note that $T(y) = TP(y)$.

For a quantum finite-state machine we have, instead:

$$\langle \psi(y)^{t+p}| = \langle \psi^t | U^p P(y) \ . \tag{12}$$

In both cases we dropped the renormalization factor.

The stochastic language generated by a particular quantum finite-state generator G for a particular measurement period p is labeled $\mathcal{L}^p(G)$. Consider now the set of languages generated by G for varying measurement period $\{\mathcal{L}^p(G)\}$.

Definition. *The* language diversity *of a (quantum or classical) finite-state machine* G *is the logarithm of the total number* $|\{\mathcal{L}^p(G)\}|$ *of stochastic languages that* G *generates as a function of measurement period* p*:*

$$\mathcal{D}(G) = \log_2 |\{\mathcal{L}^p(G)\}| \ . \tag{13}$$

Whenever we are interested in comparing the diversity in terms of formal and stochastic languages we add the respective subscript and write

$\mathcal{D}_{formal}(G)$ and $\mathcal{D}_{stoch}(G)$, respectively. Here, $\mathcal{D}_{formal} = log_2|\mathcal{L}^p_{formal}|$. In general, $\mathcal{D}_{stoch}(G) > \mathcal{D}_{formal}(G)$ for any particular G.

In the following we will demonstrate several properties related to the language diversity of classical and quantum finite-state machines.

Since every $\mathcal{L}(QDG)$ is generated by some DG, at first blush one might conclude that DGs are at least as powerful as QDGs. However, as pointed out in Ref. [7], this is true only for one particular measurement period. In the following examples we will study the dependence of the generated languages on the measurement period. It will become clear that Theorem 1 does not capture all of the properties of a QDG and its classical analog DG. For all but the periodic processes of the following examples the *language diversity* is larger for the QDG than its DG analog, even though the projection operators are identical.

These observations suggest the following.

Conjecture. $\mathcal{D}(QDG) \geq \mathcal{D}(DG)$.

The inequality becomes an equality in one case.

Proposition 1. *For a QDG G generating a periodic stochastic language $\mathcal{L}$ and its analog DG G'*

$$\mathcal{D}(G) = \mathcal{D}(G') \ . \tag{14}$$

Proof. *For any measurement period p and word length L words $y^L \in \mathcal{L}(G)$ and $y'^L \in \mathcal{L}(G')$ with $y^L = y'^L$ have the same probability:* $\Pr(y^L) = \Pr(y'^L)$. *That is,*

$$\Pr(y^L) = \|\psi^0 U^p P(y_0) U^p P(y_1) \ldots U^p P(y_{L-1})\|^2$$

and

$$\Pr(y'^L) = \langle \pi^0 | T^p P(y_0) T^p P(y_1) \ldots T^p P(y_{L-1}) | \eta \rangle \ .$$

Due to determinism and periodicity $\Pr(y^L) = 0$ *or* 1, *and also* $\Pr(y'^L) = 0$ *or* 1 *for all possible* ψ^0 *and* π^0, *respectively. Since* $U = T$, *the probabilities are equal.* □

We can give an upper bound for $\mathcal{D}$ in this case.

Proposition 2. *For a QG G generating a periodic process language $\mathcal{L}$ with period N:*

$$\mathcal{D}(G) \leq log_2(|Y| + N(N-1)) \ . \tag{15}$$

Proof. *Since $\mathcal{L}(G)$ is periodic, $\mathcal{L}^p(G) = \mathcal{L}^{p+N}(G)$. For $p = N, 2N, \ldots$: $\mathcal{L}^p(G) = \{y^*\}$, $y \in Y$. For $p = N+i, 2N+i, \ldots$, $0 < i < N$: $\mathcal{L}^p(G) =$* $\mathrm{sub}((\sigma_0\sigma_1 \ldots \sigma_{N-1})^*)$ *and all its cyclic permutations are generated, in total N for each p. This establishes an upper bound of $|Y| + N(N-1)$.*

For general quantum processes there exists an upper bound for the language diversity.

Proposition 3. *For a QGD G*

$$\mathcal{D}(G) \leq log_2(|Y| + k(k-1)) \ , \quad (16)$$

where k is the integer giving

$$U^k = I + \iota J \ , \quad (17)$$

I is the identity matrix, $\iota \ll 1$, *and J is a diagonal matrix* $\sum_i |J_{ii}|^2 \leq 1$.

Proof. *It was shown in Ref. [6] (Thms. 6 and 7), that any* $n \times n$ *unitary U can be considered as rotating an* $n-$*dimensional torus. Then for some k* U^k *is within a small distance of the identity matrix. Thus, k can be considered the* pseudo-period *of the process, compared to a strictly periodic process with* period *N and* $U^N = I$.

Thus, $\mathcal{L}^p(G)$ *and* $\mathcal{L}^{p+k}(G)$ *are* δ*-similar with* $\delta \ll 1$. *For* $p = k$: $U^p = I + \iota J$, *generating* $\mathcal{L} = \{y^*\}$. *Using the same argument as in the proof of Prop. 2 to lower the bound by k this establishes the upper bound for* $\mathcal{D}(G)$.□

It should be noted that the upper bound on $\mathcal{D}$ depends on the parameter δ defining the similarity of languages $\mathcal{L}^p(G)$ and $\mathcal{L}^{p+k}(G)$. In general, the smaller δ is, the larger is k.

Proposition 4. *For a QDG G generating a periodic process language the number of formal languages* $|\mathcal{L}_{formal}(G)|$ *equals the number of stochastic languages* $|\mathcal{L}_{stoch}(G)|$

$$\mathcal{D}_{formal}(G) = \mathcal{D}_{stoch}(G). \quad (18)$$

Proof. *It is easily seen that any QG generating a periodic process is deterministic: its unitary matrix has only* 0 *and* 1 *entries. It follows that word probabilities are either* 0 *or* 1 *and so there is a one-to-one mapping between the stochastic language generated and the corresponding formal language.*□

Corollary 1. *For a QDG G generating a periodic process and its analog DG G′:*

$$\mathcal{D}_{formal}(G) = \mathcal{D}_{formal}(G') = \mathcal{D}_{stoch}(G) = \mathcal{D}_{stoch}(G') \ . \quad (19)$$

Proof. *The Corollary follows from Prop. 1 and a straightforward extension of Proposition 4 to classical periodic processes.*□

6 Examples

The first two examples, the iterated beam splitter and the quantum kicked top, are quantum dynamical systems that are observed using complete measurements. In quantum mechanics, a *complete measurement* is defined as a nondegenerate measurement operator, i.e., one with nondegenerate eigenvalues. The third example, the distinct period-5 processes, illustrates processes observed via incomplete measurements. Deterministic quantum and stochastic finite-state generators are constructed and compared for each example.

6.1 Iterated beam splitter

The iterated beam splitter is a simple quantum process, consisting of a photon that repeatedly passes through a loop of beam splitters and detectors, with one detector between each pair of beam splitters [7]. Thus, as the photon traverses between one beam splitter and the next, its location in the upper or lower path between them is measured nondestructively by the detectors. The resulting output sequence consists of symbols 0 (upper path) and 1 (lower path).

The operators have the following matrix representation in the experiment's eigenbasis:

$$U = \frac{1}{\sqrt{2}} \begin{pmatrix} 1 & 1 \\ 1 & -1 \end{pmatrix} , \tag{20}$$

$$P(0) = \begin{pmatrix} 1 & 0 \\ 0 & 0 \end{pmatrix} , \tag{21}$$

$$P(1) = \begin{pmatrix} 0 & 0 \\ 0 & 1 \end{pmatrix} . \tag{22}$$

Observing with different measurement periods, the generated language varies substantially. As can be easily seen with Eqs. (10) and (12), three (and only three) languages are generated as one varies p. They are summarized in Table 1 for all $y^L \in \mathcal{L}$ and for $n = 0, 1, 2 \ldots$, which is used to parametrize the measurement period. The language diversity of the QDG is then $\mathcal{D} = \log_2(3)$. We can compare this to the upper bound given in Prop. 3. In the case of the unitary matrix U given above $k = 2$, since $UU = I$. U is also known as the *Hadamard* matrix. Thus, the upper bound for the language diversity in this case is $\mathcal{D} \leq log_2(4)$.

The classical equivalent DG for the iterated beam splitter, constructed as described in Ref. [7], is given by the following state-to-state transition matrix:

Iterated Beam Splitter Language Diversity				
Machine Type	p	$\text{supp}(\mathcal{L})$	$\mathcal{L}$	$\mathcal{D}$
QDG	$2n$	$(0+1)^*$	$\Pr(y^L) = 2^{-L}$	
	$2n+1$	0^*	$\Pr(y^L) = 1$	
	$2n+1$	1^*	$\Pr(y^L) = 1$	1.58
DG	n	$(0+1)^*$	$\Pr(y^L) = 2^{-L}$	0

Table 1. Process languages generated by the *QDG* for the iterated beam splitter and by the classical *DG*. The measurement period takes a parameter $n = 0, 1, 2 \ldots$. The word probability is given for all $y^L \in \mathcal{L}$.

$$T = \begin{pmatrix} \frac{1}{2} & \frac{1}{2} \\ \frac{1}{2} & \frac{1}{2} \end{pmatrix} .$$

Using Eqs. (5) and (11), we see that only one language is generated for all p. This is the language of the *fair coin* process, a random sequence of 0s and 1s, see Table 1. Thus, $\mathcal{D}(DG) = 0$.

6.2 Quantum kicked top

The periodically kicked top is a familiar example of a finite-dimensional quantum system whose classical limit exhibits various degrees of chaotic behavior as a function of its control parameters [12]. For a spin-1/2 system the unitary matrix is:

$$U = \begin{pmatrix} \frac{1}{\sqrt{2}} & -\frac{1}{\sqrt{2}} \\ \frac{1}{\sqrt{2}} & \frac{1}{\sqrt{2}} \end{pmatrix} \cdot \begin{pmatrix} e^{-ik} & 0 \\ 0 & e^{-ik} \end{pmatrix}$$

and the projection operators are:

$$P(0) = \begin{pmatrix} 1 & 0 \\ 0 & 0 \end{pmatrix} ,$$

$$P(1) = \begin{pmatrix} 0 & 0 \\ 0 & 1 \end{pmatrix} .$$

Since this *QDG* G is deterministic, its classical *DG* G' exists and is given by:

$$T = \begin{pmatrix} \frac{1}{2} & \frac{1}{2} \\ \frac{1}{2} & \frac{1}{2} \end{pmatrix} .$$

The process languages generated by this QDG and its analog DG are given in Table 2. The language diversity is $\mathcal{D}(G) = \log_2(5)$. Whereas the language diversity of classical counterpart DG is $\mathcal{D}(G') = 0$, since it generates only the language of the fair coin process.

Spin-1/2 Quantum Kicked Top Language Diversity				
Machine Type	p	supp($\mathcal{L}$)	$\mathcal{L}$	$\mathcal{D}$
QDG	$4n+1, 4n+3$	$(0+1)^*$	$\Pr(y^L) = 2^{-L}$	
	$4n+2$	sub($(01)^*$)	$\Pr(((01)^*)^L) = 1/2$	
			$\Pr(((10)^*)^L) = 1/2$	
	$4n+2$	sub($(10)^*$)	$\Pr(((10)^*)^L) = 1/2$	
			$\Pr(((01)^*)^L) = 1/2$	
	$4n$	0^*	$\Pr(y^L) = 1$	
	$4n$	1^*	$\Pr(y^L) = 1$	2.32
DG	n	$(0+1)^*$	$\Pr(y^L) = 2^{-L}$	0

Table 2. Process languages generated by the QDG for the spin-1/2 quantum kicked top and its corresponding classical DG. The measurement period, again, is parametrized by $n = 0, 1, 2 \ldots$. The word probability is given for all $y^L \in \mathcal{L}$.

6.3 Period-5 process

As examples of periodic behavior and, in particular, of incomplete measurements, consider the binary period-5 processes distinct up to permutations and $(0 \leftrightarrow 1)$ exchange. There are only three such processes: $(11000)^*$, $(10101)^*$, and $(10000)^*$ [13]. They all have the same state-to-state transition matrix—a period-5 permutation. This irreducible, doubly stochastic matrix is responsible for the fact that the QDG of a periodic process and its classical DG have the same properties. Their state-to-state unitary transition matrix is given by

$$T = U = \begin{pmatrix} 0\,0\,1\,0\,0 \\ 0\,0\,0\,1\,0 \\ 0\,1\,0\,0\,0 \\ 0\,0\,0\,0\,1 \\ 1\,0\,0\,0\,0 \end{pmatrix} . \tag{23}$$

The projection operators differ between the processes with different template words, of course. For template word 10000, they are:

$$P(0) = \begin{pmatrix} 1&0&0&0&0 \\ 0&1&0&0&0 \\ 0&0&0&0&0 \\ 0&0&0&1&0 \\ 0&0&0&0&1 \end{pmatrix}, \tag{24}$$

$$P(1) = \begin{pmatrix} 0&0&0&0&0 \\ 0&0&0&0&0 \\ 0&0&1&0&0 \\ 0&0&0&0&0 \\ 0&0&0&0&0 \end{pmatrix}. \tag{25}$$

For 11000, they are:

$$P(0) = \begin{pmatrix} 1&0&0&0&0 \\ 0&0&0&0&0 \\ 0&0&0&0&0 \\ 0&0&0&1&0 \\ 0&0&0&0&1 \end{pmatrix}, \tag{26}$$

$$P(1) = \begin{pmatrix} 0&0&0&0&0 \\ 0&1&0&0&0 \\ 0&0&1&0&0 \\ 0&0&0&0&0 \\ 0&0&0&0&0 \end{pmatrix}. \tag{27}$$

And for word 10101, they are:

$$P(0) = \begin{pmatrix} 0&0&0&0&0 \\ 0&1&0&0&0 \\ 0&0&0&0&0 \\ 0&0&0&0&0 \\ 0&0&0&0&1 \end{pmatrix}, \tag{28}$$

$$P(1) = \begin{pmatrix} 1&0&0&0&0 \\ 0&0&0&0&0 \\ 0&0&1&0&0 \\ 0&0&0&1&0 \\ 0&0&0&0&0 \end{pmatrix}. \tag{29}$$

The difference between the measurement alphabet size and the period of a process, which determines the number of states of a periodic process, should be noted. In all our examples the measurement alphabet is binary.

Thus, in having five internal states but only a two-letter measurement alphabet, the period-5 processes necessarily constitute systems observed via incomplete measurements.

The set of languages generated by the three processes is summarized in Table 3. The generated language depends on the initial state only when the measurement period is a multiple of the process period.

Distinct Period-5 Processes' Language Diversity				
Machine Type	p	supp($\mathcal{L}$)	$\mathcal{L}, L > 5$	$\mathcal{D}$
10000	$5n+1, 5n+2$ $5n+3, 5n+4$	sub($(10000)^*$)	$\Pr(y^L) = 1/5$	
	$5n$	0^*	$\Pr(y^L) = 1$	
	$5n$	1^*	$\Pr(y^L) = 1$	1.58
11000	$5n+1, 5n+4$	sub($(11000)^*$)	$\Pr(y^L) = 1/5$	
	$5n+2, 5n+3$	sub($(01010)^*$)	$\Pr(y^L) = 1/5$	
	$5n$	0^*	$\Pr(y^L) = 1$	
	$5n$	1^*	$\Pr(y^L) = 1$	2
10101	$5n+1, 5n+4$	sub($(10101)^*$)	$\Pr(y^L) = 1/5$	
	$5n+2, 5n+3$	sub($(00111)^*$)	$\Pr(y^L) = 1/5$	
	$5n$	0^*	$\Pr(y^L) = 1$	
	$5n$	1^*	$\Pr(y^L) = 1$	2

Table 3. Process languages produced by the three distinct period-5 generators. The quantum and classical versions are identical in each case. The measurement period is parametrized by $n = 0, 1, 2 \ldots$. For simplicity, the word probability is given for all $y^L \in \mathcal{L}$ with $L \geq 5$. For the nontrivial languages above, when $L > 5$ there are only five words at each length, each having equal probability.

The language diversity for the process 10000 is $\mathcal{D} = \log_2(3)$ and for both the processes 11000 and 10101, $\mathcal{D} = 2$. Note that the processes 11000 and 10101 generate each other at particular measurement periods, if one exchanges 0s and 1s. It is not surprising therefore that the two models have the same language diversity.

It turns out that the state of the quantum systems under periodic dynamics is independent of the measurement protocol. At each point in time the system is in an eigenstate of the measurement operator. Therefore, the measurement does not alter the internal state of the quantum system. Thus, a system in state $\langle\psi_0|$ is going to be in a particular state $\langle\psi_2|$ after two time steps, independent of whether being measured in between. This is true for quantum and classical periodic systems. The conclusion is that for periodic processes there is no difference between unmeasured quantum and classical states. This is worth noting, since this is the circumstance where classical and quantum systems are supposed to differ. As a consequence the language diversity is the same for the quantum and classical model of all periodic processes, which coincides with Prop. 1.

Note, however, that the language diversity is not the same for all processes with the same period. A property that is reminiscent of the transient information [9, 13], which also distinguishes between structurally different periodic processes.

6.4 Discussion

The examples show that the language diversity monitors aspects of a process's structure and it is different for quantum and classical models of aperiodic processes. This suggests that it will be a useful aid in discovering structure in the behavior of quantum dynamical systems. For the aperiodic examples, the QDG had a larger language diversity than its classical DG. And this suggests a kind of computational power of QDGs that is not obvious from the structural constraints of the machines. Language diversity could be compensation, though, for other limitations of QDGs, such as not being able to generate all regular languages. The practical consequences of this for designing quantum devices remains to be explored.

A comparison between QDGs and their classical DGs gives a first hint at the structure of the lowest levels of a potential hierarchy of quantum computational model classes. It turned out that for periodic processes a QDG has no advantage over a DG in terms of the diversity of languages possibly generated by any QDG. However, for the above examples of both incomplete and complete measurements, the set of generated stochastic languages is larger for a QDG than the corresponding DG.

Table 4 summarizes the processes discussed above, their properties and language diversities. All finite-state machines are deterministic, for which case it was shown that there exists an equivalent DG that generates the same language [7]. This is true, though only for one particular measurement period. Here we expanded on those results in comparing a range of measurement periods and the entire set of generated stochastic languages.

For each example quantum generator and the corresponding classical generator the language diversity and the type of measurement (complete/incomplete) are given. For all examples the language diversity is larger for the QDG than the DG. It should be noted, however, that the

	Quantum process	Classical process
System	Iterated beam splitter	Fair coin
$\mathcal{D}$	$\log_2(3)$	0
Measurement	Complete	Complete
System	Quantum kicked top	Fair coin
$\mathcal{D}$	$log_2(5)$	0
Measurement	Complete	Complete
System	10000	10000
$\mathcal{D}$	$\log_2(3)$	$\log_2(3)$
Measurement	Incomplete	Incomplete
System	11000	11000
$\mathcal{D}$	2	2
Measurement	Incomplete	Incomplete
System	10101	10101
$\mathcal{D}$	2	2
Measurement	Incomplete	Incomplete

Table 4. Comparison between QDGs and their classical DGs. Note that the term "(in)complete measurement" is not used for classical systems. However, the above formalism does render it meaningful. It is used in the same way as in the quantum case (one-dimensional subspaces or non-degenerate eigenvalues).

fair coin process is also generated by a one-state DG with transition matrices $T(0) = T(1) = (1/2)$. This it not true for the QDGs. Thus, the higher language diversity of a QDG is obtained at some cost—a larger number of states is needed than with a DG generating any one particular process language. The situation is different, again, for the period-5 processes—there is no DG with fewer states that generates the same process language.

The above examples were simple in the sense that their language diversity is a finite, small number. In some broader sense, this means that they are recurrent—to use terminology from quantum mechanics. For other processes the situation might not be quite as straightforward. To find the language diversity one has to take the limit of large measurement periods. For implementations this is a trade-off, since larger measurement period requires a coherent state for a longer time interval. In particular it should be noted that in the above examples shorter intervals between measurements cause more "interesting" observed behavior. That is, the stochastic language $\mathcal{L}^2 = \{(01)^*, (10)^*\}$ generated by the quantum kicked top with $\Pr(y^L) = 1/2$, consisting of strings with alternating 0s and 1s is more structured than the language $\mathcal{L}^4 = \{0^*\}$ with $\Pr(y^L) = 1$ consisting of only 0s. (Cf. Table 2.)

7 Conclusion

Quantum finite-state machines occupy the lowest level of an as-yet only partially known hierarchy of quantum computation. Nonetheless, they are useful models for quantum systems that current experiment can implement, given the present state of the art. We briefly reviewed quantum finite-state generators and their classical counterparts—stochastic finite-state generators. Illustrating our view of computation as an intrinsic property of a dynamical system, we showed similarities and differences between finite-memory classical and quantum processes and, more generally, their computational model classes. In particular, we introduced the language diversity—a new property that goes beyond the usual comparison of classical and quantum machines. It captures the fact that, when varying measurement protocols, different languages are generated by quantum systems. Language diversity appears when quantum interference operates.

For a set of examples we showed that a deterministic quantum finite-state generator has a larger language diversity than its classical analog. Since we associate a language with a particular behavior, we also associate a set of languages with a set of possible behaviors. As a consequence, the *QDG*s all exhibited a larger set of behaviors than their classical analogs. That is, they have a larger capacity to store and process information.

We close by suggesting that the design of finite quantum computational elements could benefit from considering the measurement process not only as a final but also as an intermediate step, which may simplify experimental design.

Since we considered only finite-memory systems here, their implementation is already feasible with current technology. Cascading compositions of finite processes can rapidly lead to quite sophisticated behaviors, as discussed in Ref. [7]. A discussion of associated information storage and processing capacity analogous to those used for classical dynamical systems in Ref. [9] is under way.

Acknowledgments

Partial support was provided by DARPA Agreement F30602-00-2-0583. KW's postdoctoral fellowship was provided by the Wenner-Gren Foundations, Stockholm, Sweden.

References

1. R. P. Feynman, Simulating physics with computers. *International Journal of Theoretical Physics*, 21:467–488, 1982.
2. E. Knill, R. Laflamme, R. Martinez, and C.-H. Tseng, An algorithmic benchmark for quantum information processing. *Nature*, 405:368–370, 2002.
3. P. Shor, Polynomial-time algorithms for prime factorization and discrete logarithms on a quantum computer. In *Proceedings of the 35th Annual Symposium on Foundations of Computer Science*, page 124, 1994. e-print arxiv/quant-ph/9508027.
4. J. E. Hopcroft, R. Motwani, and J. D. Ullman, *Introduction to Automata Theory, Languages, and Computation.* Addison-Wesley, 2001.
5. A. Kondacs and J. Watrous, On the power of quantum finite state automata. In *38th IEEE Conference on Foundations of Computer Science*, pages 66–75, 1997.
6. C. Moore and J. P. Crutchfield, Quantum automata and quantum grammars. *Theoretical Computer Science*, 237:275–306, 2000.
7. K. Wiesner and J. P. Crutchfield, Computation in finitary quantum processes. *in preparation*, 2006.
8. A. Bertoni and M. Carpentieri, Analogies and differences between quantum and stochastic automata. *Theoretical Computer Science*, 262:69–81, 2001.
9. J. P. Crutchfield and D. P. Feldman, Regularities unseen, randomness observed: Levels of entropy convergence. *Chaos*, 13:25 – 54, 2003.
10. A. Paz, *Introduction to Probabilistic Automata.* New York Academic Press, 1971.
11. C. Howson, Theories of probability. *The British Journal for the Philosophy of Science*, 46:1–32, 1995.
12. F. Haake, M. Kús, and R. Scharf, Classical and quantum chaos for a kicked top. *Z. Phys. B*, 65:381–395, 1987.
13. D. P. Feldman and J. P. Crutchfield, Synchronizing to periodicity: The transient information and synchronization time of periodic sequences. *Advances in Complex Systems*, 7(3-4):329–355, 2004.

Logic circuits in a system of repelling particles

William M. Stevens

Department of Physics and Astronomy, Open University, Walton Hall,
Milton Keynes, MK7 6AA, UK
william@stevens93.fsnet.co.uk,
WWW home page: http://www.srm.org.uk

Abstract. A model of a kinematic system consisting of moveable tiles in a two dimensional discrete space environment is presented. Neighbouring tiles repel one another, some tiles are fixed in place. A dual-rail logic gate is constructed in this system.

1 Introduction

This paper is motivated by a desire to understand how computing devices can be built from simple mechanical parts. An attempt is made to model the kinematic behaviour of a simple system without modelling any mechanism that gives rise to the kinematic behaviour in question. The model used can be classified as a 'kinematic automaton' and can be simulated using a cellular automaton.

This scheme arose during research into automatic construction and self-replication. Several researchers have proposed or built programmable constructing machines capable of automatically building a wide range of other machines under program control from a collection of prefabricated parts ([8], [5], [7]). If one is interested in simplifying the process of automatically constructing a complex device, then using a small range of component part types is advantageous because the complexity of the constructing device is likely to be lower than if a large range of part types were used.

Several abstract models of self-replicating systems based around programmable constructors have been devised ([8], [2], [11]), and at least one simple physical self-replicating system of this kind has been built ([12]). One problem that arises when considering how to make physical programmable self-replicators is that of devising a control system using the range of prefabricated parts that are available to the replicator. If some of the prefabricated parts have a built-in capability for processing digital information (as in [11]), or even a built-in computer (as in [12]), the problem is solvable, but the solution is not ideal because the complexity of such parts tends to be fairly high.

Some of the systems cited above are based around sets of parts that contain both mechanical elements and digital-processing elements, other systems are based on complex parts that combine both functions. It may be possible to reduce the range and/or the complexity of the set of parts required to build a self-replicating constructor by processing digital information using the mechanical interactions between parts and ultimately using a mechanical computer as the control unit for a programmable constructor. This paper does not go that far, but does show how a logic gate can be made using simple mechanical interactions and a small range of part types.

This result is also of interest as a collision-based computing scheme. In [4], Fredkin and Toffoli showed that elastic collisions between idealised billiard balls (and fixed mirrors) obeying Newton's laws of motion can be used to implement boolean logic elements. Fredkin and Toffoli were interested in showing that reversible, conservative logic is physically possible. This paper is not concerned with conservative logic, but with logic elements made from simple parts that are technologically plausible. In [6] Margolus showed how to model Fredkin and Toffoli's system using a cellular automaton. Collision-based computing with a simple basis has also been demontrated in several other cellular automata environments ([9], [3]). While these environments are mathematically simple they are not designed to model physical behaviour.

2 A simple kinematic simulation environment

A two dimensional discrete space, discrete time simulation environment that supports moveable square tiles is studied in this paper. The behaviour of a tile is very simple: neighbouring tiles repel one another, but some tiles can be fixed in place. Figure 1 illustrates the behaviour of tiles, and should be read to mean that any occurrence of the configuration to the left an arrow will evolve into the configuration to the right of the arrow after one time step. All other configurations remain unchanged.

This behaviour can be specified more formally using the set of cellular automaton rules given in Fig. 2.

These rules specify a cellular automaton with an eight cell neighbourhood and three states per cell. A cell can be empty, or it can contain a fixed tile, or it can contain a moveable tile. In Fig. 2 empty cells are denoted by a square with a dashed boundary, fixed tiles are denoted by

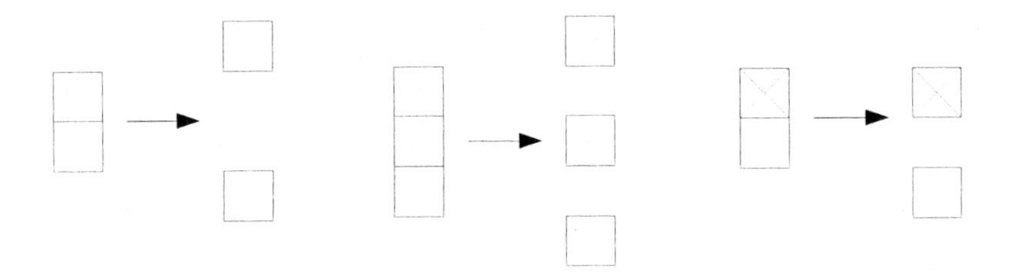

Fig. 1. Three rules that completely describe how tiles interact

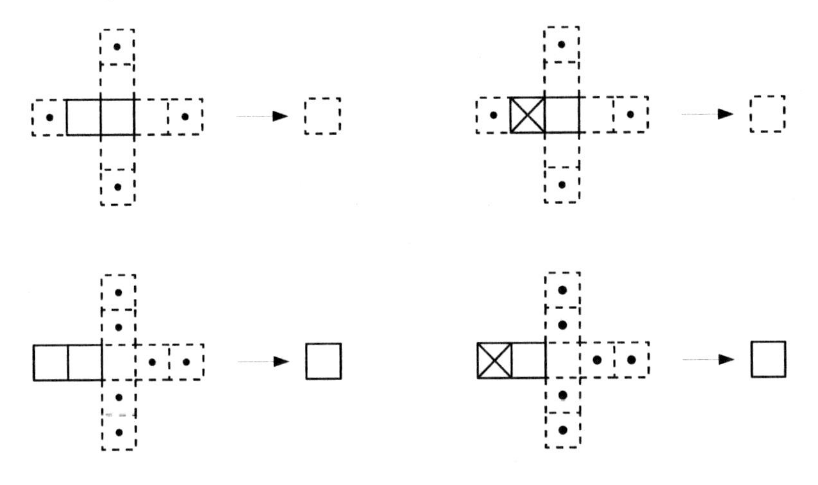

Fig. 2. Tile behaviour specified as a set of cellular automaton rules

a square with a cross in, moveable tiles are denoted by a square with a solid boundary. Cells with a dot in the centre can be in any of the three states. The rules are symmetrical, so if a configuration of tiles rotated through any multiple of 90 degrees matches a rule, the central cell changes to the state to the right of the arrow on the next time step. If a configuration matches none of these rules, the central cell remains in the same state.

These rules are sufficient for modelling all of the mechanisms described in this paper.

3 Some basic mechanisms

It is relatively straightforward to devise configurations of tiles that behave as 'one-shot' logic devices, in which arrangements of tiles compute a function once but cannot be re-used after the computation is finished. This paper shows how re-usable circuits can be constructed which are capable of carrying out one computation after another.

Figures 3 to 11 show nine basic mechanisms from which more complex mechanisms can be put together.

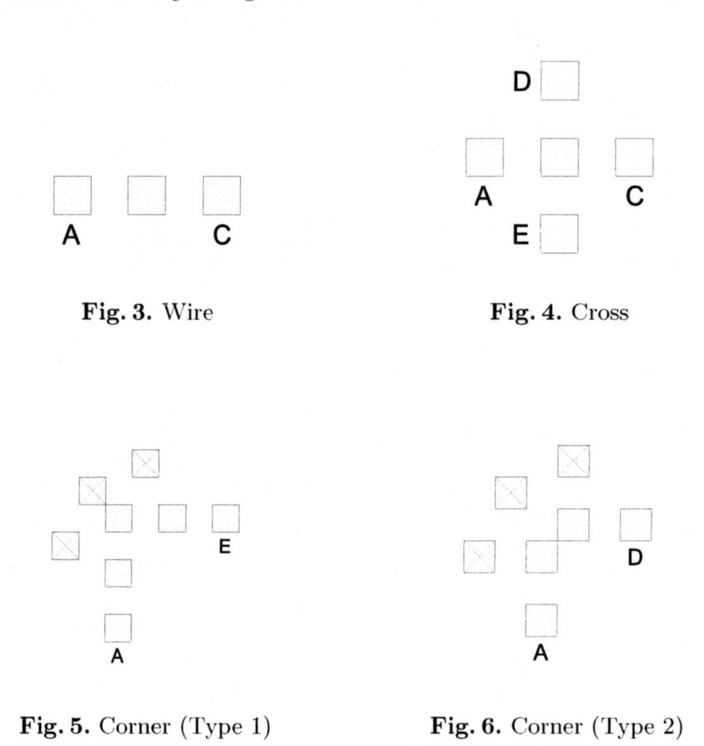

Fig. 3. Wire

Fig. 4. Cross

Fig. 5. Corner (Type 1)

Fig. 6. Corner (Type 2)

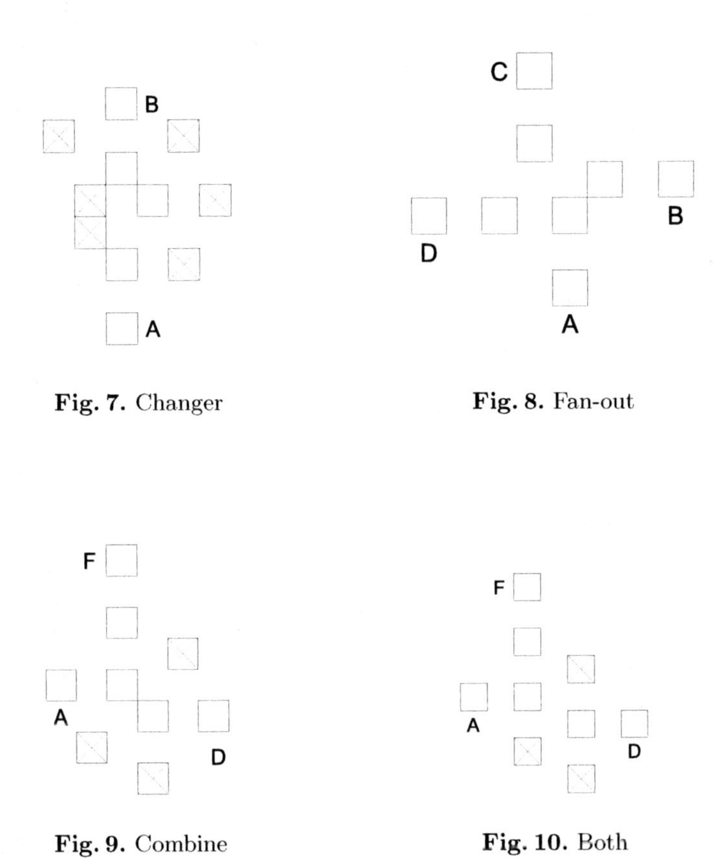

Fig. 7. Changer

Fig. 8. Fan-out

Fig. 9. Combine

Fig. 10. Both

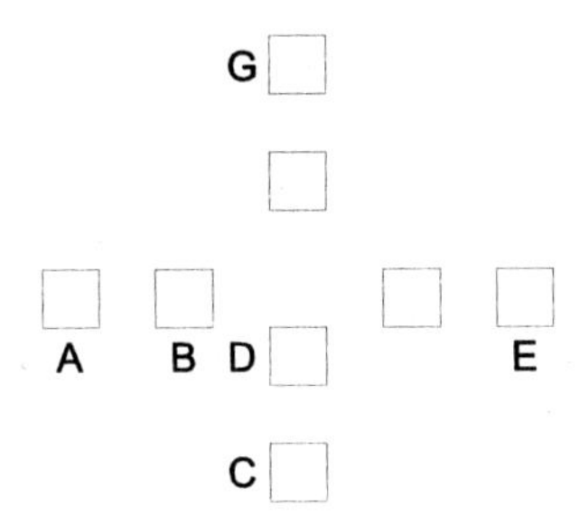

Fig. 11. Hold

The action of each of these mechanisms can be described formally. The notation used is intuitive and will be introduced as it is used. The letters n,e,s and w that appear in square brackets in equations denote the directions north (up the page), south (down the page), east (to the right of the page) and west (to the left of the page).

3.1 Wire

For the Wire in Figure 3 we can write:

$$A[e] \begin{array}{l} \xrightarrow{1} A[w] \\ \xrightarrow{2} C[e] \end{array} \tag{1}$$

To mean that the result of displacing A eastward by one unit at time t_A is that A will be displaced westward at $t_A + 1$, and C will be displaced eastward at $t_A + 2$. The centre tile in the Wire will also move, but will return to its original position. A Wire can thus be thought of as a path along which a signal can propagate, where the signal consists of a displacement of a tile away from its normal position in the path. Wires of any length can be made.

Note that a Wire also works in reverse, so we can also write:

$$C[w] \begin{array}{l} \xrightarrow{1} C[e] \\ \xrightarrow{2} A[w] \end{array} \tag{2}$$

If both ends of a Wire are moved at once, the result is that two signals cancel each other out as follows:

$$A[e],C[w] \begin{array}{l} \xrightarrow{1} A[w] \\ \xrightarrow{1} C[e] \end{array} \tag{3}$$

This 'cancelling out' behaviour is used extensively by the logic gate described in section 5.

3.2 Cross

For the Cross in Fig. 4 we can write:

$$A[e] \xrightarrow{1} A[w], \quad A[e] \xrightarrow{2} C[e] \tag{4}$$

$$C[w] \xrightarrow{1} C[e], \quad C[w] \xrightarrow{2} A[w] \tag{5}$$

$$D[s] \xrightarrow{1} D[n], \quad D[s] \xrightarrow{2} E[s] \tag{6}$$

$$E[n] \xrightarrow{1} E[s], \quad E[n] \xrightarrow{2} D[n] \tag{7}$$

$$A[e],C[w] \xrightarrow{1} A[w], \quad A[e],C[w] \xrightarrow{1} C[e] \tag{8}$$

$$D[s],E[n] \xrightarrow{1} D[n], \quad D[s],E[n] \xrightarrow{1} E[s] \tag{9}$$

Note that the Cross mechanism misbehaves when $(A[e] \vee C[w]) \wedge (D[s] \vee E[n])$ at any time t, so any circuit that uses the Cross mechanism must avoid this situation.

3.3 Corner (Type 1)

For the Type 1 Corner in Fig. 5 we can write:

$$A[n] \xrightarrow{1} A[s], \quad A[n] \xrightarrow{4} E[e] \tag{10}$$

$$E[w] \xrightarrow{1} E[e], \quad E[w] \xrightarrow{4} A[s] \tag{11}$$

3.4 Corner (Type 2)

For the Type 2 Corner in Fig. 6 we can write:

$$A[n] \xrightarrow{1} A[s], \quad A[n] \xrightarrow{3} D[e] \tag{12}$$

$$D[w] \xrightarrow{1} D[e], \quad D[w] \xrightarrow{3} A[s] \tag{13}$$

Note that both types of Corner can propagate signals in either direction. If we only need a Corner to work in one direction then one of its fixed tiles can be removed.

3.5 Changer

For the Changer in Fig. 7 we can write:

$$A[n]\ \begin{array}{l} \nearrow^{1} A[s] \\ \xrightarrow{7} B[n] \end{array} \tag{14}$$

The Changer is so-called because it alters the spacing of the tiles in a signal path. This is often necessary when joining mechanisms together.

3.6 Fanout

For the Fanout mechanism in Fig. 8 we can write:

$$A[n]\ \begin{array}{l} \nearrow^{1} A[s] \\ \nearrow^{3} D[e] \\ \xrightarrow{4} F[n] \\ \searrow_{5} H[w] \end{array} \tag{15}$$

If we need fewer than 3 outputs, any of the output paths in the Fanout mechanism can be replaced with a single fixed tile.

3.7 Combine

For the Combine mechanism in Fig. 9 we can write:

$$A[e]\ \begin{array}{l} \nearrow^{1} A[w] \\ \xrightarrow{5} F[n] \end{array} \tag{16}$$

$$D[w]\ \begin{array}{l} \nearrow^{1} D[e] \\ \xrightarrow{4} F[n] \end{array} \tag{17}$$

The Combine mechanism also has useful behaviour when driven from tile F:

$$F[s]\ \begin{array}{l} \nearrow^{1} F[n] \\ \xrightarrow{5} A[w] \\ \searrow_{4} D[e] \end{array} \tag{18}$$

3.8 Both

The Both mechanism will produce an output only after both of its inputs have been stimulated:

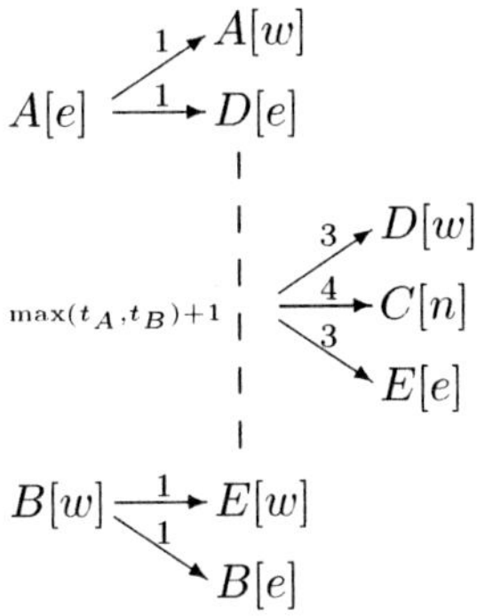

$$
\begin{array}{lll}
 & \nearrow^{1} & A[w] \\
A[e] & \xrightarrow{1} & D[e] \\
 & | & \\
 & | & \nearrow^{3}\ D[w] \\
\max(t_A, t_B)+1 & | & \xrightarrow{4}\ C[n] \\
 & | & \searrow_{3}\ E[e] \\
 & | & \\
B[w] & \xrightarrow{1} & E[w] \\
 & \searrow_{1} & B[e]
\end{array}
\tag{19}
$$

In equation 19 the dashed line with arrows leading to subsequent events indicates that the occurrence of two events (at time $\max(t_A, t_B)+1$) causes the subsequent events.

Because tiles D and E return to their original positions during the operation of the Both mechanism its behaviour can be described more concisely:

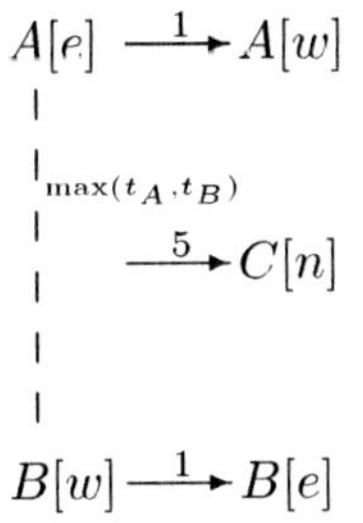

$$
\begin{array}{lll}
A[e] & \xrightarrow{1} & A[w] \\
| & & \\
|_{\max(t_A, t_B)} & & \\
| & \xrightarrow{5} & C[n] \\
| & & \\
| & & \\
B[w] & \xrightarrow{1} & B[e]
\end{array}
\tag{20}
$$

3.9 Hold

The Hold mechanism in Fig. 11 consists of four paths meeting at a junction, and is used as follows. At time t_A tile A may or may not be displaced eastward. At time t_E tile E may or may not be displaced westward. So that at time $t_m = \max(t_A, t_E) + 1$ the Hold mechanism may be in one of the four possible states shown in Fig. 11 to 14. At time $t_C \geq t_m$ tile C is displaced northward, and the response of the mechanism depends upon which of the four states it is in. Let us call the arrangements shown in Figs. 11, 12, 13 and 14 H_0, H_1, H_2 and H_3 respectively.

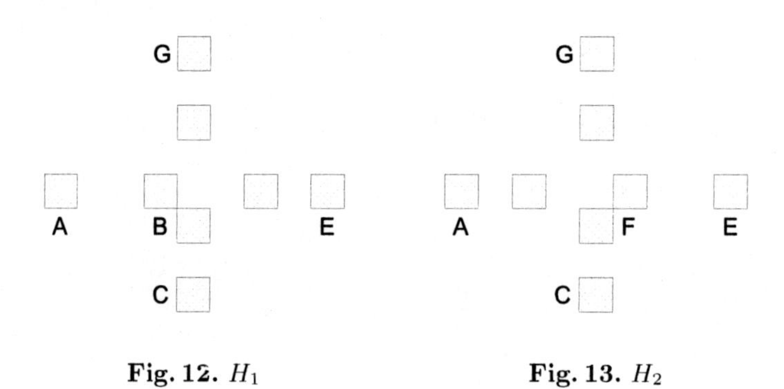

Fig. 12. H_1 **Fig. 13.** H_2

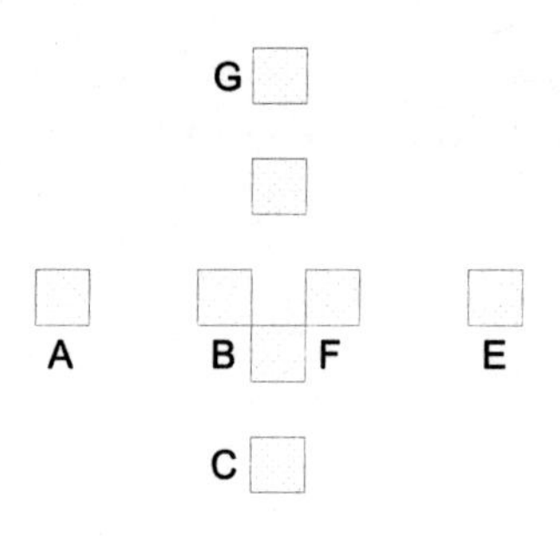

Fig. 14. H_3

Of H_0 we can say:

$$C[n] \begin{array}{l} \xrightarrow{1} C[s] \\ \xrightarrow{1} D[n] \end{array} \tag{21}$$

Of H_1 we can say:

$$C[n] \begin{array}{l} \xrightarrow{1} C[s] \\ \xrightarrow{2} B[w] \\ \xrightarrow{4} E[e] \end{array} \tag{22}$$

Of H_2 we can say:

$$C[n] \begin{array}{l} \xrightarrow{1} C[s] \\ \xrightarrow{2} F[e] \\ \xrightarrow{4} A[w] \end{array} \tag{23}$$

And of H_3 we can say:

$$C[n] \begin{array}{l} \xrightarrow{1} C[s] \\ \xrightarrow{2} B[w] \\ \xrightarrow{2} F[e] \end{array} \tag{24}$$

What we have in effect is a mechanism that allows us to collide two signals (entering at A and E) together without having to worry about the relative timing of the two signals, because the collision only takes place when a signal is applied at C. When the Hold mechanism is used in a circuit, any signals emerging from A or E after the collision will propagate away from the Hold mechanism, leaving it in the state shown in Fig. 15, which we call H_4. We can return the mechanism to its original state H_0 by applying a signal at G.

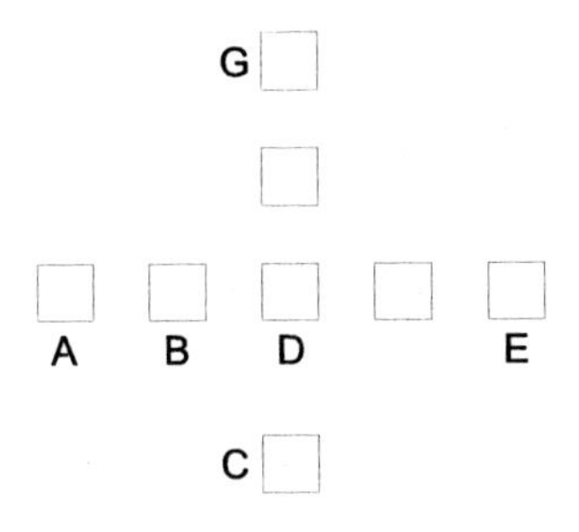

Fig. 15. H_4

For H_4:

$$G[s] \xrightarrow{2} D[s], \quad G[s] \xrightarrow{1} G[n] \tag{25}$$

Table 1 shows the logical operations that are effected by the collision of two signal paths in a Hold mechanism.

Inputs before t_C	Outputs after t_C
Neither A nor E	Neither A nor E
A but not E	E but not A
E but not A	A but not E
Both A and E	Neither A nor E

Table 1. A collision between two signal paths

The fact that a signal will only be output at E if a signal is input at A but not at E before time t_C forms the basis of the logic gate described in section 5.

4 Circuits

Circuits can be made by connecting mechanisms together, and in the next section a dual-rail logic gate is made using the nine mechanisms described in the previous section. Before doing this, it is necessary to show how to describe the behaviour of two mechanisms joined to one another in such a way that a signal emerging from one will enter another.

Figure 16 shows a Wire and a Combine mechanism that have tile B in common. We saw earlier how to describe the behaviour of each of these mechanisms individually, and using this knowledge we can write:

$$A[e] \xrightarrow{1} A[w], \quad A[e] \xrightarrow{2} B[e]; \quad B[e] \xrightarrow{1} B[w], \quad B[e] \xrightarrow{5} C[s], \quad B[e] \xrightarrow{4} D[n] \tag{26}$$

In words, this means that a displacement of A eastward by one unit at t_A causes A to be displaced westward at $t_A + 1$ and B to be displaced eastward at $t_A + 2$. The displacement of B subsequently causes B to

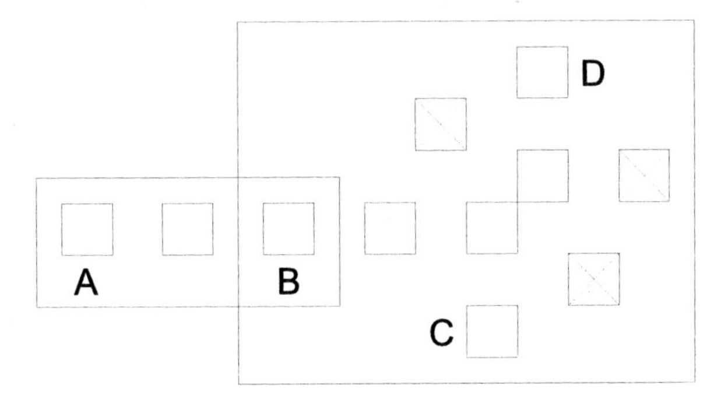

Fig. 16. Two mechanisms joined via tile B

be displaced westward at $t_A + 2 + 1$, C to be displaced southward at $t_A + 2 + 5$ and D to be displaced northward at $t_A + 2 + 4$.

Because there is no net movement of B, we can shorten 26 to:

$$A[e] \begin{array}{l} \overset{1}{\nearrow} A[w] \\ \overset{7}{\longrightarrow} C[s] \\ \underset{6}{\searrow} D[n] \end{array} \tag{27}$$

5 A dual-rail logic gate

In subsection 3.9 we saw how the Hold mechanism effects a logical operation. In order to use this logical operation as the basis for a logic gate, additional circuitry is needed.

The A and E inputs to the Hold mechanism also serve as outputs after a signal has been applied at C. To enable us to extract a signal returning along a path which is also used as an input we can use the mechanism shown in Fig. 17, called a 'uni-directional gate with tap'. An analysis of this mechanism is given in equations 28 and 29.

$$\begin{array}{cccccccccccccc} & \overset{1}{\nearrow} A[s] & & \overset{1}{\nearrow} B[e] & & \overset{1}{\nearrow} C[s] & & \overset{1}{\nearrow} D[s] & & \overset{1}{\nearrow} E[w] & & \overset{1}{\nearrow} F[w] & & \overset{1}{\nearrow} G[n] \\ A[n] & \xrightarrow{4} & B[w] & \xrightarrow{6} & C[n] & \xrightarrow{2} & D[n] & \xrightarrow{3} & E[e] & \xrightarrow{2} & F[e] & \xrightarrow{5} & G[s] & \xrightarrow{5} I[e] \end{array} \tag{28}$$

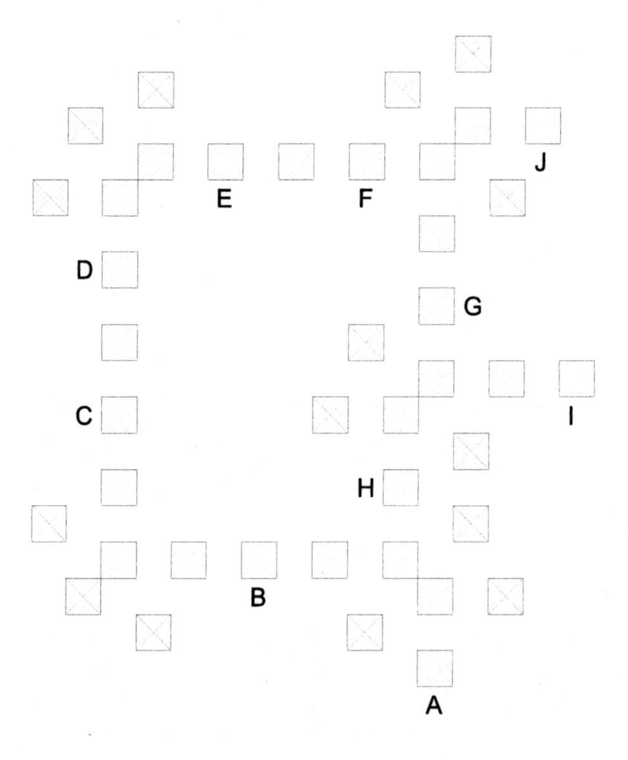

Fig. 17. Uni-directional gate with tap

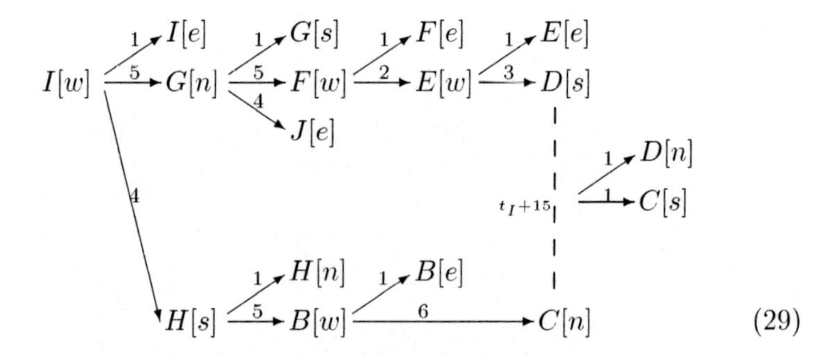

(29)

In equation 29 (as in equation 19) the dashed line with arrows leading to subsequent events indicates that the occurrence of two events (at time $t_I + 15$ in this case) causes the subsequent behaviour (in this case, the behaviour of a wire when both ends are displaced at once).

By observing which tiles return to their original positions during the course of the operation of this mechanism, we can shorten these equations to:

$$A[n] \begin{array}{l} \nearrow^{1} A[s] \\ \xrightarrow{27} I[e] \end{array} \tag{30}$$

$$I[w] \begin{array}{l} \nearrow^{1} I[e] \\ \xrightarrow{9} J[e] \end{array} \tag{31}$$

Thus, by applying a signal to A we can inject a signal onto a path connected to I. If a signal enters the mechanism at I, it will emerge at J.

Note that the number of tiles used in this mechanism could be reduced by shortening some of the paths. However, if we were to do this then we would not be able to analyse the mechanism in terms of the nine basic mechanisms described previously. To simplify the analysis and to make it clear which of the basic mechanisms we are using, we will use longer paths than are necessary.

A derivative of the mechanism in Fig. 17 is shown in Fig. 18. This is called a 'uni-directional gate'. The behaviour of the uni-directional gate is given in equations 32 and 33.

$$A[n] \begin{array}{l} \nearrow^{1} A[s] \\ \xrightarrow{27} I[e] \end{array} \tag{32}$$

$$I[w] \xrightarrow{1} I[e] \tag{33}$$

Thus, the uni-directional gate will permit a signal to travel from A to I, but not in the reverse direction.

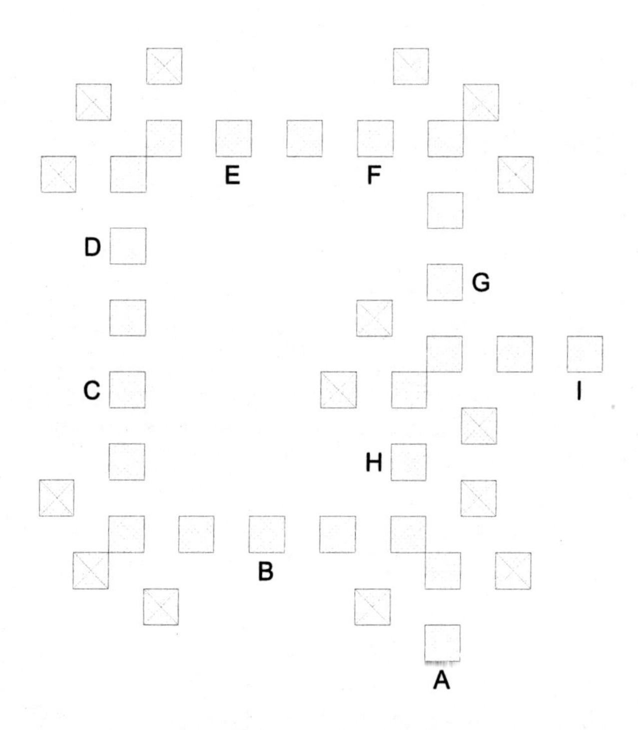

Fig. 18. Uni-directional gate

Recall that in dual-rail or 1-of-2 logic, every logical value X is represented by a pair of binary values X_0 and X_1 so that $X \leftrightarrow (X_0 = 0 \wedge X_1 = 1)$ and $\bar{X} \leftrightarrow (X_0 = 1 \wedge X_1 = 0)$. We can implement a dual-rail logic scheme by using two signal paths corresponding to X_0 and X_1 in such a way that the passage of a signal along one path represents logic 0, and the passage of a signal along the other path represents logic 1. For a dual-rail logic gate with two logical inputs A and B (and therefore four signal path inputs A_0, A_1, B_0 and B_1), it is possible to detect when both logical inputs have been received using the mechanism shown in Fig. 19.

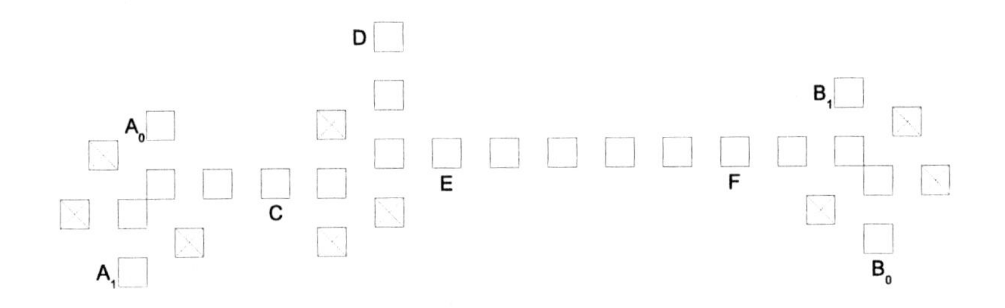

Fig. 19. Dual-rail input detection mechanism

There are 4 possible cases to analyse for this mechanism :

A	B	A_0	A_1	B_0	B_1	Equation
False	False	1	0	1	0	35
False	True	1	0	0	1	36
True	False	0	1	1	0	37
True	True	0	1	0	1	38

A complete analysis for the first case in this table is given in equation 34, which can be shortened to equation 35. Complete analyses for the other three cases are omitted for the sake of brevity since they follow a similar pattern to equation 34.

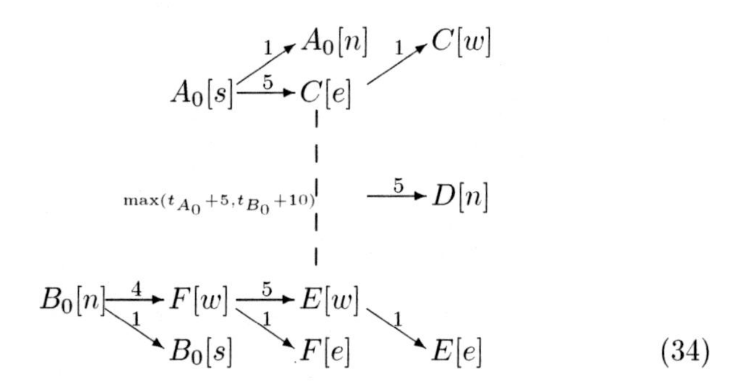

(34)

$$\begin{array}{l} A_0[s] \xrightarrow{1} A_0[n] \\ \max(t_{A_0}, t_{B_0}+4) \\ \quad \xrightarrow{10} D[n] \\ B_0[s] \xrightarrow{1} B_0[n] \end{array} \qquad (35)$$

$$\begin{array}{l} A_0[s] \xrightarrow{1} A_0[n] \\ \max(t_{A_0}, t_{B_1}+5) \\ \quad \xrightarrow{10} D[n] \\ B_1[s] \xrightarrow{1} B_1[n] \end{array} \qquad (36)$$

$$\begin{array}{l} A_1[s] \xrightarrow{1} A_1[n] \\ \max(t_{A_1}, t_{B_0}+5) \\ \quad \xrightarrow{9} D[n] \\ B_0[s] \xrightarrow{1} B_0[n] \end{array} \qquad (37)$$

$$\begin{array}{l} A_1[s] \xrightarrow{1} A_1[n] \\ \max(t_{A_1}, t_{B_1}+5) \\ \quad \xrightarrow{9} D[n] \\ B_1[s] \xrightarrow{1} B_1[n] \end{array} \qquad (38)$$

Equations 35 to 38 show that a signal will emerge at D for for every possible combination of logical values that A and B can take.

When describing the Hold mechanism shown in Fig. 11 we noted that it will be used by firstly applying signals at A or E (or both, or neither), then applying a signal at C, and afterwards applying a signal at D to reset the mechanism. The augmented hold mechanism shown in Fig. 20 allows us to dispense with having to apply a reset signal by deriving a reset signal from the signal at C.

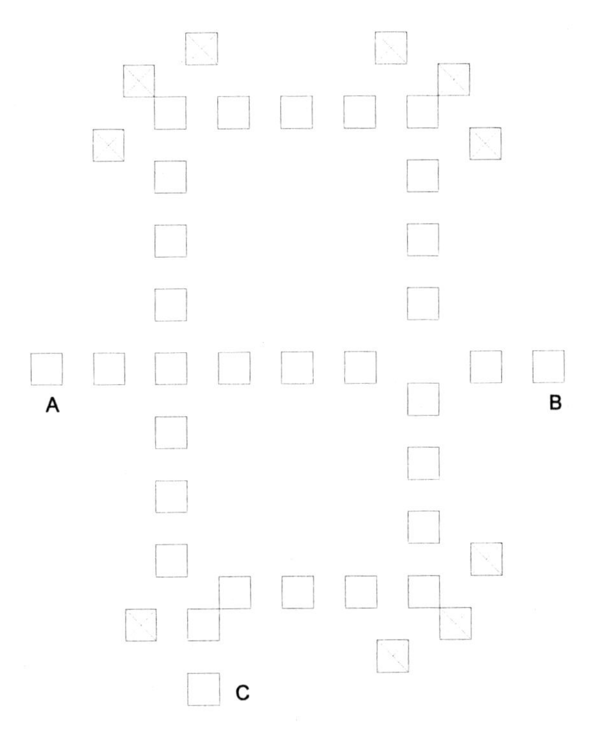

Fig. 20. Augmented hold mechanism

The net behaviour of the augmented hold mechanism is given in equations 39 to 42 for the four possible cases that result if signals are applied at A or B or both or neither.

$$\begin{array}{l} A[e] \xrightarrow{1} A[w] \\ \vdots \\ \vdots_{\max(t_A, t_C+5)} \\ \vdots \quad \xrightarrow{8} B[e] \\ \vdots \\ C[n] \xrightarrow{1} C[s] \end{array} \qquad (39)$$

$$\begin{array}{l} B[w] \xrightarrow{1} B[e] \\ \vdots \\ \vdots_{\max(t_B, t_C+5)} \\ \vdots \quad \xrightarrow{8} A[w] \\ \vdots \\ C[n] \xrightarrow{1} C[s] \end{array} \qquad (40)$$

$$A[w] \xrightarrow{1} A[e] \qquad\qquad C[n] \xrightarrow{1} C[s] \qquad (42)$$

$$B[w] \xrightarrow{1} B[e]$$

$$C[n] \xrightarrow{1} C[s] \qquad (41)$$

Equation 41 holds so long as $t_C \geq \max(t_A - 5, t_B - 9)$.

Using a uni-directional gate with tap, a uni-directional gate, an augmented hold mechanism, a dual-rail input detection mechanism, two Combine mechanisms, a Changer and some connecting wires we can make the mechanism shown in Fig. 21. This mechanism feeds signals derived from A_0 and B_1 into an augmented hold mechanism, where a collision will take place once the dual-rail input detect mechanism indicates that all required inputs have been received. After the collision a signal will emerge at O if a signal was applied at A_0 but not at B_1. A signal derived from the output of the dual-rail input detect mechanism will emerge at P regardless of the logical values of A and B.

Equation 43 shows an analysis of the case for $A_0[e]$ at time t_{A_0} and $B_0[n]$ at time t_{B_0}.

$$A_0[e] \xrightarrow{1} A_0[w], \quad A_0[e] \xrightarrow{5} K[n], \quad A_0[e] \xrightarrow{4} C[s]$$

$$K[n] \xrightarrow{1} K[s], \quad K[n] \xrightarrow{28} L[e] \xrightarrow{1} L[w]$$

$$C[s] \xrightarrow{8} D[s] \xrightarrow{1} D[n], \quad C[s] \xrightarrow{1} C[n]$$

$$\xrightarrow{8} M[e] \quad \max(t_{A_0}+8, t_{B_0})+27$$

$$M[e] \xrightarrow{1} M[w], \quad M[e] \xrightarrow{1} N[e], \quad N[e] \xrightarrow{1} N[w], \quad N[e] \xrightarrow{9} O[w]$$

$$\max(t_{A_0}+8, t_{B_0})+4 \quad \xrightarrow{10} E[n]$$

$$E[n] \xrightarrow{8} F[n], \quad E[n] \xrightarrow{1} E[s], \quad F[n] \xrightarrow{12} P[w], \quad F[n] \xrightarrow{1} F[s]$$

$$B_0[n] \xrightarrow{1} B_0[s] \qquad (43)$$

This can be simplified to equation 44. Equations for the other three cases are shown in 45 to 47. (Complete analyses for these are not given since they follow a similar pattern to 43).

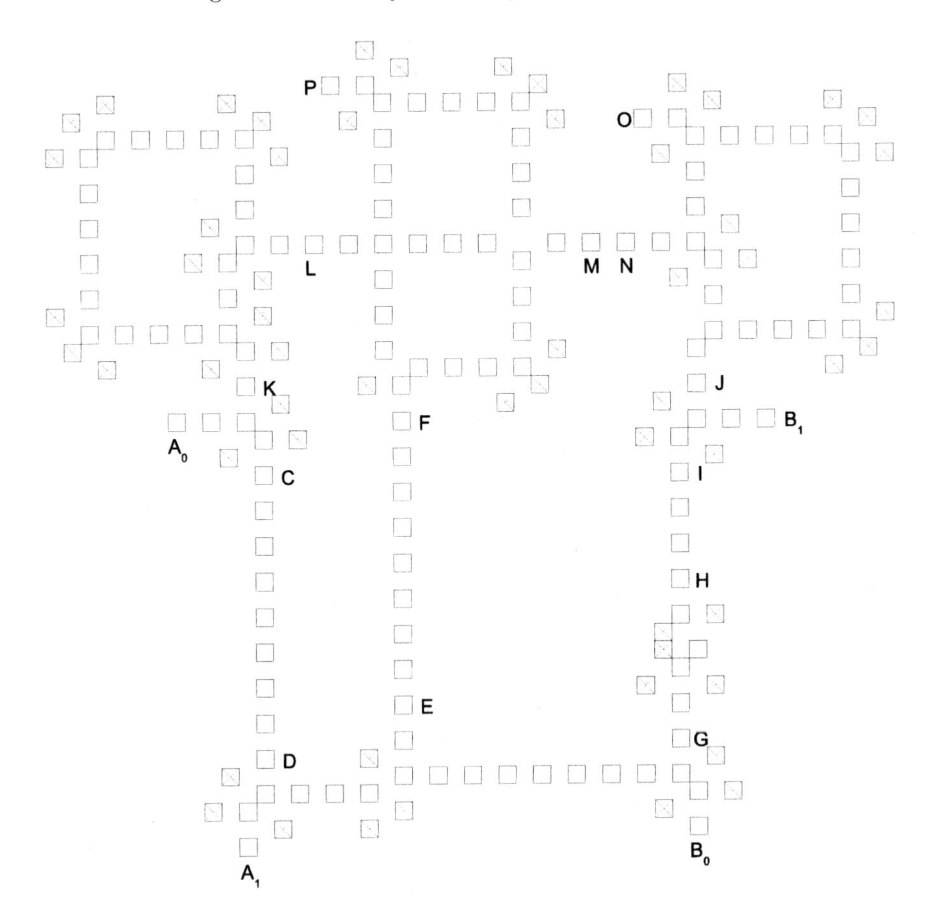

Fig. 21. Logic gate - Stage 1

$$
\begin{array}{l}
A_0[e] \xrightarrow{1} A_0[w] \\
\quad \vdots \; {}_{\max(t_{A_0}+8,\, t_{B_0})+33} \\
\qquad \xrightarrow{1} P[w] \\
\qquad \searrow^{12}\; O[w] \\
\quad \vdots \\
B_0[n] \xrightarrow{1} B_0[s]
\end{array}
\qquad (44)
$$

$$
\begin{array}{l}
A_0[e] \xrightarrow{1} A_0[w] \\
\quad \vdots \; {}_{\max(t_{A_0},\, t_{B_1}+7)+42} \\
\qquad \xrightarrow{1} P[w] \\
\quad \vdots \\
B_1[w] \xrightarrow{1} B_1[e]
\end{array}
\qquad (45)
$$

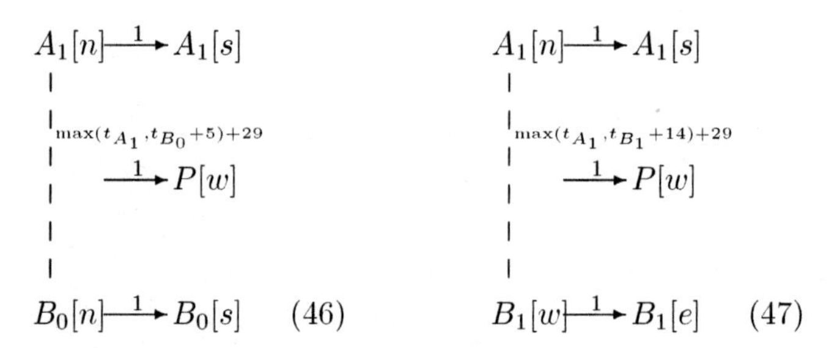

Thus in all four cases, a signal will emerge at P.

In the case shown in equation 44 (where signals are applied at A_0 and B_0), a signal will emerge at P at time $t_P = \max(t_{A_0} + 8, t_{B_0}) + 34$ and another signal will emerge at O at time $t_O = \max(t_{A_0} + 8, t_{B_0}) + 45 = t_P + 11$.

The mechanism shown in Fig. 22 can be used to derive a signal from O to use as the Q_1 output of a dual-rail logic gate, and to derive a signal from P and O to use as the Q_0 output of a dual-rail logic gate.

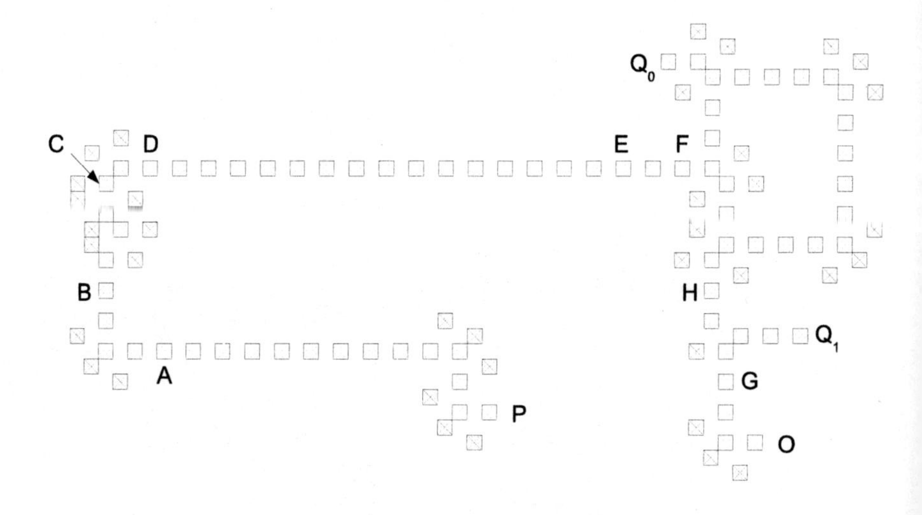

Fig. 22. Logic gate - Stage 2

An analysis of this mechanism is given in equations 48 and 49.

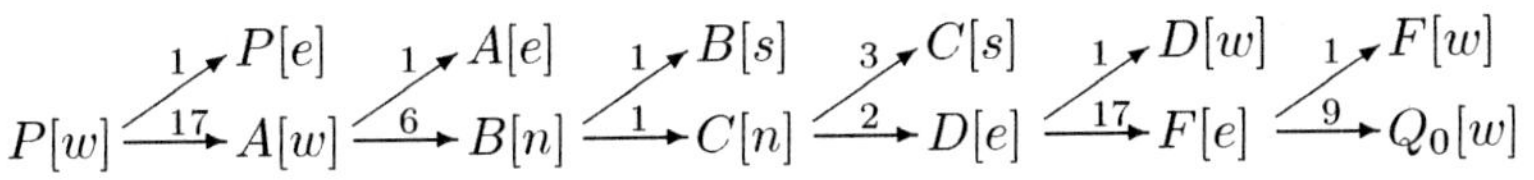

$$\begin{array}{lllllll}
 & {}^{1}\!\nearrow P[e] & {}^{1}\!\nearrow A[e] & {}^{1}\!\nearrow B[s] & {}^{3}\!\nearrow C[s] & {}^{1}\!\nearrow D[w] & {}^{1}\!\nearrow F[w] \\
P[w] & \xrightarrow{17} A[w] & \xrightarrow{6} B[n] & \xrightarrow{1} C[n] & \xrightarrow{2} D[e] & \xrightarrow{17} F[e] & \xrightarrow{9} Q_0[w]
\end{array} \tag{48}$$

$$\begin{array}{lllllll}
 & {}^{1}\!\nearrow P[e] & {}^{1}\!\nearrow A[e] & {}^{1}\!\nearrow B[s] & {}^{1}\!\nearrow C[s] & {}^{1}\!\nearrow D[w] & \\
P[w] & \xrightarrow{17} A[w] & \xrightarrow{6} B[n] & \xrightarrow{7} C[n] & \xrightarrow{2} D[e] & \xrightarrow{17} E[e] & \\
 & & & & & \vdots & \nearrow E[w] \\
 & & & & & \vdots & \xrightarrow[1]{1} F[e] \\
 & & & & {}^{4}\!\nearrow Q_1[e] & \vdots & \\
 & & O[w] & \xrightarrow{1} G[n] & \xrightarrow{4} H[n] & \xrightarrow{27} F[w] & \\
 & & & {}_{5}\!\searrow O[e] & {}_{1}\!\nwarrow G[s] & {}_{1}\!\nwarrow H[s] &
\end{array} \tag{49}$$

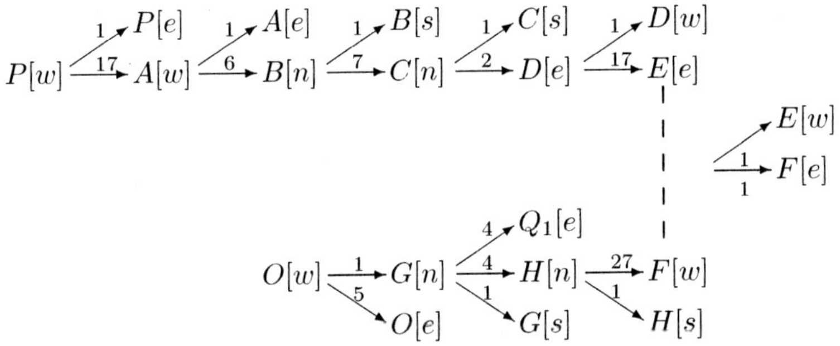

These equations can be simplified to 50 and 51 and used in conjunction with 44 to 47 to deduce the overall behaviour of the logic gate shown in Fig. 23.

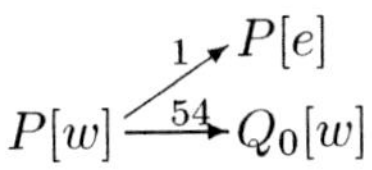

$$\begin{array}{ll}
 & {}^{1}\!\nearrow P[e] \\
P[w] & \xrightarrow{54} Q_0[w]
\end{array} \tag{50}$$

$$\begin{array}{ll}
P[w] & \xrightarrow{1} P[e] \\
\vdots & \\
\vdots_{\,t_P+49} & \\
\vdots & \\
 & {}^{9}\!\nearrow Q_1[e] \\
O[w] & \xrightarrow{1} O[e]
\end{array} \tag{51}$$

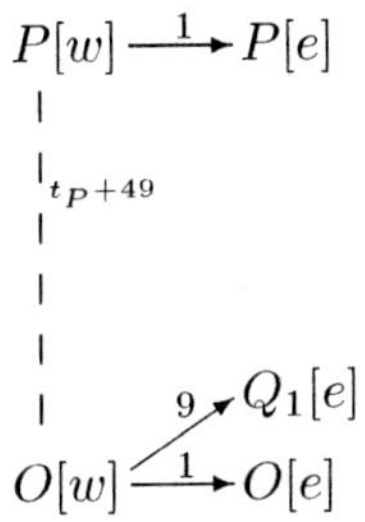

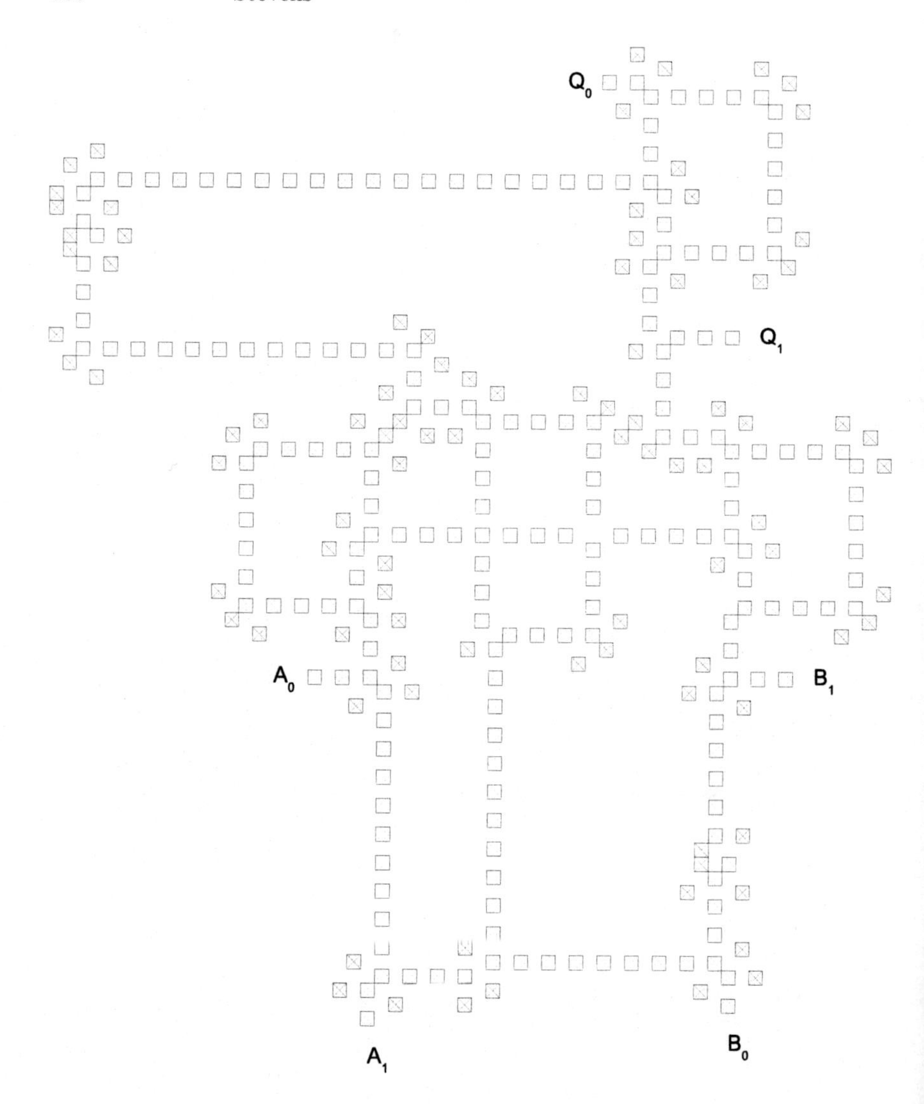

Fig. 23. A dual-rail NAND gate

Thus, the overall behaviour of our dual-rail logic gate is described by table 2. From this, it can be seen the the logic gate in Fig. 23 is a dual-rail implementation of a boolean NAND gate.

A	B	Inputs to gate	Outputs from gate	Q
False	False	$A_0[e]$ at t_{A_0} , $B_0[n]$ at t_{B_0}	$Q_1[e]$ at $\max(t_{A_0}+8, t_{B_0})+92$	True
False	True	$A_0[e]$ at t_{A_0} , $B_1[w]$ at t_{B_1}	$Q_0[e]$ at $\max(t_{A_0}, t_{B_1}+7)+97$	False
True	False	$A_1[n]$ at t_{A_1} , $B_0[n]$ at t_{B_0}	$Q_0[e]$ at $\max(t_{A_1}, t_{B_0}+5)+84$	False
True	True	$A_1[n]$ at t_{A_1} , $B_1[w]$ at t_{B_1}	$Q_0[e]$ at $\max(t_{A_1}, t_{B_1}+14)+84$	False

Table 2. Logical behaviour of the gate in Fig. 23

6 Conclusion

The number of tiles in Fig. 23 could be reduced by shortening the paths between mechanisms and by reducing the size of some mechanisms which were deliberately kept larger than necessary in order to clarify their structure. Unused fixed tiles could also be removed at some corners and in some Combine mechanisms.

The part of the logic gate that performs the logic operation is the augmented hold mechanism. If we placed constraints on signal timing at the inputs to the gate, we could do without this mechanism and replace it with a wire, resulting in a simpler gate. However, if we were to do this and then attempt to connect several logic gates together to make a circuit we would have to introduce delays between one gate and another in order to meet timing constraints.

Many tiles in the logic gate are involved with separating signals travelling in one direction along a path from signals travelling in the opposite direction. Note that in Fredkin and Toffoli's Billiard Ball model and in some logic schemes based on glider collisions in two and three dimensionsal cellular automata (for example [10]) signal separation of this kind is not necessary because in these environments particles can be given a velocity component perpendicular to the collision axis, so that the results from a collision automatically end up in a different location from the 'inputs' to the collision.

One aim of this work was to find a simple and technologically plausible basis for computating using a small range of simple kinematical part types. To assess whether the system described meets this aim, further work is needed. Are there any physical systems of repelling particles in which the system described can be implemented?

One problem that may need to be addressed in a physical system with classical behaviour is that of emulating a synchronously-updating discrete grid in an asynchronous continuous system. One approach may be to arrange things so that the substrate on which particles move lies closely parallel to a regular array of attractors that can be switched on and off periodically, and to which particles are attracted so strongly that the repulsion between neighbouring particles can be overcome. When the attractors are switched on, the particles will align themselves with the regular array. When the attractors are switched off, the particles can interact and move.

Another approach may be to adapt the system described here so that global synchronization of the system is not required. Adachi et al ([1]) have shown that some asynchronous cellular automata are capable of supporting universal computation. Further work is needed to determine whether the CA rules specified in Fig. 2 support universal computation if used in an ACA model.

Such speculations cannot proceed far without deeper research into the physics of repelling particles in various different physical environments.

Software to simulate the system described in this paper can be obtained at http://www.srm.org.uk

References

1. Adachi, S., Peper, F., Lee, Jia., Computation by asynchronously updating cellular automata. J. Stat. Phys. **114(1/2)** (2004) 261–289
2. Codd, E.F., Cellular Automata. Academic Press, New York (1968)
3. Cook, M., Universality in elementary cellular automata. Complex Systems **15** (2004) 1–40
4. Fredkin, E., Toffoli, T., Conservative logic. J. Theo. Phys. (1982) 219–253
5. Freitas, R., Gilbreath, W. P., Advanced automation for space missions. Section 5.6.4: Robot replication feasibility demonstration. NASA Conference Publication CP-2255 (1982) 253–257
6. Margolus, N., Physics-like models of computation. Physica D **10** 1984 81-95
7. Moses, M., A Physical Prototype of a Self-Replicating Universal Constructor. Master Thesis, Department of Mechanical Engineering, University of New Mexico (2001). http://www.home.earthlink.net/~mmoses152/SelfRep.doc
8. Von Neumann, F., Theory of Self-Reproducing Automata. Edited and completed by A.W. Burks. University of Illinois Press, Urbana Illinois (1966) 81–82
9. Rendell, P., Turing universality of the Game of Life. In: Adamatzky A., Editor, Collision-Based Computing (Springer, 2002) 513–539.
10. Rennard, J., Implementation of logical functions in the Game of Life. In: Adamatzky A., Editor, Collision-Based Computing (Springer, 2002) 491–512
11. Stevens, W.M., A programmable constructor in a kinematic environment. Poster presentation at Micro and Nanotechnology 2005 conference. http://www.srm.org.uk/papers/CBlocks3D_programmable_constructor.pdf
12. Zykov, V., Mytilinaios, E., Adams, B., Lipson, H., Self-reproducing machines. Nature **435(7038)** (2005) 163–164.

From group theory to reversible computers

Alexis De Vos and Yvan Van Rentergem

Imec v.z.w. and Universiteit Gent, B-9000 Gent, Belgium

Abstract. Reversible logic circuits of a certain logic width form a group, isomorphic to a symmetric group. Its Young subgroups allow systematic synthesis of an arbitrary reversible circuit. We can choose either a left coset, right coset, or double coset approach. The synthesis methods are beneficial to both classical and quantum computer design. As an illustration, three experimental prototypes of reversible computing devices are presented.

1 Introduction

Reversible computing [1] [2] is useful both in lossless classical computing [3] [4] and in quantum computing [5]. It can be implemented in both classical and quantum hardware technologies. In the present paper, we will demonstrate the application of group theory to the detailed design.

Reversible logic circuits distinguish themselves from arbitrary logic circuits by two properties:

- the number of output bits always equals the number of input bits and
- for each pair of different input words, the two corresponding output words are different.

For instance, it is clear that an `AND` gate is not reversible, as

- it has only one output bit, but two input bits and
- for three different input words, the three corresponding output words are equal.

Table 1a, on the other hand, gives an example of a reversible circuit. Here, the number of inputs equals the number of outputs, i.e. two. This number is called the width w of the reversible circuit. The table gives all possible input words AB. We see how all the corresponding output

Table 1. Truth table of three reversible logic circuits of width 2: (a) an arbitrary reversible circuit r, (b) the identity gate i, and (c) the inverse r^{-1} of r.

AB	PQ
0 0	0 0
0 1	1 0
1 0	1 1
1 1	0 1

(a)

AB	PQ
0 0	0 0
0 1	0 1
1 0	1 0
1 1	1 1

(b)

AB	PQ
0 0	0 0
0 1	1 1
1 0	0 1
1 1	1 0

(c)

words PQ are different. Therefore, in contrast to arbitrary logic circuits, reversible logic circuits form a group. Table 1b gives the identity gate i and Table 1c gives r^{-1}, i.e. the inverse of r. The reader will easily verify that not only the cascade rr^{-1}, but also the cascade $r^{-1}r$ equals i.

In the present paper, we will apply different techniques from group theory to the synthesis and the analysis of reversible circuits. In Section 2, we give a brief overview of those group theoretical tools we will be using throughout the article. Sections 3, 4, and 5 discuss cosets and double cosets. They provide the most original contributions of the article. Finally, in Section 6, we discuss the relationship with some experimental prototypes, presented in earlier publications [6] [7] [8].

2 Group theory

All reversible circuits of the same width form a group. If we denote by w the width, then the truth table of an arbitrary reversible circuit has 2^w rows. As all output words have to be different, they can merely be a repetition of the input words in a different order. In other words: the 2^w output words are a permutation of the 2^w input words. There exist $(2^w)!$ ways to permute 2^w objects. Therefore there exist exactly $(2^w)!$ different reversible logic circuits of width w. The number 2^w is called the degree of the group; the number $(2^w)!$ is called the order of the group. The group is isomorphic to a group well-known by mathematicians: the symmetric group $\mathbf{S}_{2^w}$. For further reading in the area of symmetric groups in particular, and groups, subgroups, cosets, and double cosets in general, the reader is referred to appropriate textbooks [9] [10].

Table 2. Truth table of four reversible logic circuits of width 3: (a) a conservative circuit, (b) a linear circuit, (c) a univariate circuit, and (d) an exchanging circuit.

ABC	PQR
0 0 0	0 0 0
0 0 1	0 0 1
0 1 0	1 0 0
0 1 1	1 0 1
1 0 0	0 1 0
1 0 1	1 1 0
1 1 0	0 1 1
1 1 1	1 1 1

(a)

ABC	PQR
0 0 0	1 0 0
0 0 1	0 0 0
0 1 0	0 0 1
0 1 1	1 0 1
1 0 0	1 1 1
1 0 1	0 1 1
1 1 0	0 1 0
1 1 1	1 1 0

(b)

ABC	PQR
0 0 0	1 0 0
0 0 1	1 0 1
0 1 0	0 0 0
0 1 1	0 0 1
1 0 0	1 1 0
1 0 1	1 1 1
1 1 0	0 1 0
1 1 1	0 1 1

(c)

ABC	PQR
0 0 0	0 0 0
0 0 1	0 0 1
0 1 0	1 0 0
0 1 1	1 0 1
1 0 0	0 1 0
1 0 1	0 1 1
1 1 0	1 1 0
1 1 1	1 1 1

(d)

The symmetric group has a wealth of properties. For example, it has a lot of subgroups, of which most have been studied in detail. Some of these subgroups naturally make their appearance in the study of reversible computing. An example is the subgroup of conservative logic circuits, studied in detail by Fredkin and Toffoli [11]. Table 2a gives an example. In each of its rows, the output $(P, Q, R, ...)$ contains a number of 1s equal to the number of 1s in the corresponding input $(A, B, C, ...)$. An even more important subgroup is the subgroup of linear reversible circuits. Linear reversible circuits have been studied in detail by Patel et al. [12]. They play an important role in the synthesis of stabilizer circuits [13]. A logic circuit is linear iff each of its outputs P, Q, ... is a linear function of the inputs A, B, ... In its turn, a linear function is defined as follows. A Boolean function $f(A, B, ...)$ is linear iff its Reed–Muller expansion [14] only contains terms of degree 0 and terms of degree 1. The reversible circuit of Table 2a is not linear. Indeed it can be written as a set of three Boolean equations:

$$\begin{aligned} P &= B \oplus AB \oplus AC \\ Q &= A \\ R &= C \oplus AB \oplus AC \; . \end{aligned}$$

Whereas the function $Q(A, B, C)$ is linear, the function $P(A, B, C)$ is clearly not (its Reed–Muller expansion containing two terms of second degree). Table 2b is an example of a linear circuit:

$$\begin{aligned} P &= 1 \oplus B \oplus C \\ Q &= A \\ R &= A \oplus B \; . \end{aligned}$$

Based on pioneering work by Kerntopf [15], De Vos and Storme [16] have proved that an arbitrary Boolean function can be synthesized by a (loop-free and fanout-free) wiring of a finite number of identical reversible gates,reversible!gate provided the gate is not linear. In other words: all non-linear reversible circuits can be used as a universal building block. Thus the linear reversible circuits constitute the 'weak' ones. Indeed, any wiring of linear circuits (be they reversible or not, be they identical or not) can yield only linear Boolean functions at its outputs. The linear reversible circuits form a group isomorphic to what is called in mathematics the affine general linear group $AGL(w, 2)$. Its order equals $2^{(w+1)w/2}\, w!_2$, where $w!_2$ is the bifactorial of w, the q-factorial being a generalization of the ordinary factorial $w! = w!_1$:

$$w!_q = 1(1+q)(1+q+q^2)...(1+q+...+q^{w-1}) \; .$$

Table 3 gives the number of different linear reversible circuits. We see that (at least for $w > 2$) a vast majority of the reversible circuits are non-linear and thus can act as universal gates.

Table 3. The number r of different reversible circuits, the number l of different linear reversible circuits, the number u of different univariate reversible circuits, and the number e of different exchanging reversible circuits, as a function of the circuit width w.

w	r	l	u	e
1	2	2	2	1
2	24	24	8	2
3	40,320	1,344	48	6
4	20,922,789,888,000	322,560	384	24

Now we can go a step further: does a finite number of copies of a single arbitrary linear reversible circuit suffice to synthesize an arbitrary given linear Boolean function? The reader will easily verify that this is not true. In order to be able to synthesize any linear Boolean function, it is necessary and sufficient that the linear building block should have at least one output that contains at least two different degree-one terms in its Reed–Muller expansion. Table 2b is such circuit, as e.g. the Reed–Muller expansion of P contains both the terms B and C (see above). Reversible Table 2c, on the contrary, is linear, but not linear-universal, as each of its outputs is a function of only one input:

$$P = 1 \oplus B$$
$$Q = A$$
$$R = C \ .$$

Circuits like Table 2c are called univariate reversible circuits. They form a subgroup of order $w!2^w$. This number is given in Table 3. The group is isomorphic to the indirect product $\mathbf{S}_w$:$\mathbf{S}_2^w$ of the symmetric group $\mathbf{S}_w$ with the wth power of the symmetric group $\mathbf{S}_2$.

If we replace the condition 'each of its outputs is a function of only one input' by 'each of its outputs equals one input', we again descend the hierarchy of subgroups. Table 2d is such a circuit:

$$P = B$$
$$Q = A$$
$$R = C\ .$$

Such circuits are called exchangers (a.k.a. swap gates). They form a subgroup isomorphic to $\mathbf{S}_w$ of order $w!$. Also this number is given in Table 3. Finally, we can impose 'each of the outputs equals the corresponding input':

$$P = A$$
$$Q = B$$
$$R = C\ .$$

This results in the trivial subgroup $\mathbf{I}$ of order 1, merely consisting of the identity gate i.

We have thus constructed a chain of subgroups:

$$\mathbf{S}_{2^w} \supset AGL(w,2) \supset \mathbf{S}_w{:}\mathbf{S}_2^w \supset \mathbf{S}_w \supset \mathbf{I}\ ,$$

with subsequent orders

$$(2^w)! > 2^{(w+1)w/2}\, w!_2 > w!2^w > w! > 1\ ,$$

where we have assumed $w > 1$. Here, the symbol $\supset$ reads 'is proper supergroup of'. For the example $w = 3$, this becomes:

$$\mathbf{S}_8 \supset AGL(3,2) \supset \mathbf{S}_3{:}\mathbf{S}_2^3 \supset \mathbf{S}_3 \supset \mathbf{I}\ ,$$

with subsequent orders

$$40{,}320 > 1{,}344 > 48 > 6 > 1\ .$$

3 Cosets

Subgroups are at the origin of a second powerful tool in group theory: cosets. If $\mathbf{H}$ (with order h) is a subgroup of the group $\mathbf{G}$ (with order g), then $\mathbf{H}$ partitions $\mathbf{G}$ into $\frac{g}{h}$ classes, all of the same size h. These equipartition classes are called cosets. We distinguish left cosets and right cosets.

The left coset of the element a of $\mathbf{G}$ is defined as all elements of $\mathbf{G}$ which can be written as a cascade ba, where b is an arbitrary element of

H. Such left coset forms an equipartition class, because of the following property: if c is member of the left coset of a, then a is member of the left coset of c. Right cosets are defined in an analogous way. Note that **H** itself is one of the left cosets of **G**, as well as one of its right cosets.

What is the reason of defining cosets? They are very handy in synthesis. Assume we want to make an arbitrary element of the group **G** in hardware. Instead of solving this problem for each of the g cases, we only synthesize the h circuits b of **H** and a single representative r_i of each other left coset ($1 \leq i \leq \frac{g}{h} - 1$). If we can make each of these $h + \frac{g}{h} - 1$ gates, we can make all the others by merely making a short cascade br_i. If we cleverly choose the subgroup **H**, we can guarantee that $h + \frac{g}{h} - 1$ is much smaller that g. We call the set of $h + \frac{g}{h} - 1$ building-blocks the library for synthesizing the g circuits of **G**.

Maslov and Dueck [17] present a method for synthesizing an arbitrary reversible circuit of width three. As a subgroup **H** of the group $\mathbf{G} = \mathbf{S}_8$, they propose all circuits with output PQR equal 000 in case of the input $ABC = 000$. This subgroup is isomorphic to $\mathbf{S}_7$. Thus the supergroup has order $g = 8! = 40,320$, whereas the subgroup has order $h = 7! = 5,040$. The subgroup partitions the supergroup into 8 cosets. Interesting is the fact, that the procedure can be repeated: for designing each of the 5,040 members of $\mathbf{S}_7$, Maslov and Dueck choose a subgroup of $\mathbf{S}_7$. They choose all reversible circuits where PQR equals 000 in case $ABC = 000$ and equals 001 in case $ABC = 001$. This is a subgroup isomorphic to $\mathbf{S}_6$ of order $6! = 720$, which partitions $\mathbf{S}_7$ into seven cosets. Etcetera. Figure 1a illustrates one step of the procedure: the 24 elements of $\mathbf{S}_4$ are fabricated by means of the 6 elements of its subgroup $\mathbf{S}_3$ plus the representatives of the 3 other cosets in which $\mathbf{S}_4$ is partitioned by $\mathbf{S}_3$. Thus Maslov and Dueck apply the following chain of subgroups:

$$\mathbf{S}_8 \supset \mathbf{S}_7 \supset \mathbf{S}_6 \supset \mathbf{S}_5 \supset \mathbf{S}_4 \supset \mathbf{S}_3 \supset \mathbf{S}_2 \supset \mathbf{S}_1 = \mathbf{I} \ , \qquad (1)$$

with subsequent orders

$$40,320 > 5,040 > 720 > 120 > 24 > 6 > 2 > 1 \ . \qquad (2)$$

They need, for synthesizing all 40,320 members of $\mathbf{S}_8$, a library of only $(7 + 6 + ... + 1) + 1 = 29$ elements (the identity gate included). For an arbitrary circuit width w, synthesis of all $(2^w)!$ members of $\mathbf{S}_{2^w}$ needs a library of $2^{2w-1} - 2^{w-1} + 1$ elements. The Maslov–Dueck subgroup chain is close to optimal, as is demonstrated in the Appendix.

Van Rentergem et al. [18] [19] also present a coset method for synthesis, however based the following subgroup **H**: all circuits from $\mathbf{G} = \mathbf{S}_8$ possessing the property $P = A$. It is isomorphic to $\mathbf{S}_4 \times \mathbf{S}_4 = \mathbf{S}_4^2$ and has order $(4!)^2 = 576$. The subgroup $\mathbf{S}_4^2$ partitions its supergroup $\mathbf{S}_8$ into

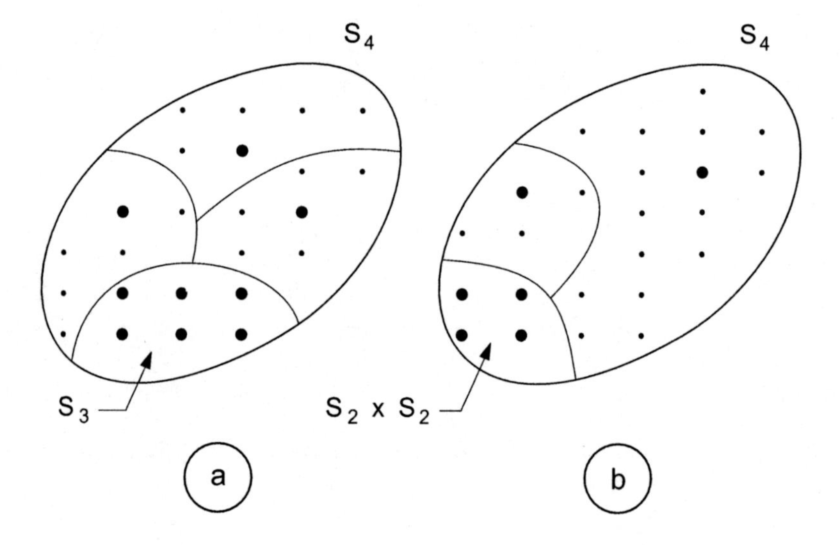

Fig. 1. The symmetric group $\mathbf{S}_4$ partitioned (a) as the four left cosets of $\mathbf{S}_3$ and (b) as the three double cosets of $\mathbf{S}_2\times\mathbf{S}_2$.
Note: the dots depict the elements of $\mathbf{S}_4$; the bold-faced dots depict the elements of the subgroup and the representatives of the (double) cosets.

five double cosets. Subsequently, the members of $\mathbf{S}_4$ are partitioned into three double cosets by use of its subgroup $\mathbf{S}_2^2$, etcetera. Thus, finally, Van Rentergem et al. apply the following chain of subgroups:

$$\mathbf{S}_8 \supset \mathbf{S}_4^2 \supset \mathbf{S}_2^4 \supset \mathbf{S}_1^8 = \mathbf{I} \ , \tag{3}$$

with subsequent orders

$$40,320 > 576 > 16 > 1 \ . \tag{4}$$

Also this chain is close to optimal. See the Appendix.

We note that the group $\mathbf{S}_{a-1}$ as well as the group $\mathbf{S}_{\frac{a}{2}} \times \mathbf{S}_{\frac{a}{2}}$ are special cases of Young subgroups of $\mathbf{S}_a$. In general, a Young subgroup [20] [21] [22] of the symmetric group $\mathbf{S}_a$ is any subgroup isomorphic to $\mathbf{S}_{a_1} \times \mathbf{S}_{a_2} \times ... \times \mathbf{S}_{a_k}$, with $(a_1, a_2, ..., a_k)$ a partition of the number a, i.e. with $a_1 + a_2 + ... + a_k = a$.

4 Double cosets

Even more powerful than cosets are double cosets. The double coset of a, element of $\mathbf{G}$, is defined as the set of all elements that can be written as b_1ab_2, where both b_1 and b_2 are members of the subgroup $\mathbf{H}$. A surprising fact is that, in general, the double cosets, in which $\mathbf{G}$ is partitioned by $\mathbf{H}$, are of different sizes (ranging from h to h^2). The number of double cosets, in which $\mathbf{G}$ is partitioned by $\mathbf{H}$, therefore is not easy to predict. It is some number between $1 + \frac{g-h}{h^2}$ and $\frac{g}{h}$. Usually, the number is much smaller than $\frac{g}{h}$, leading to the (appreciated) fact that there are far fewer double cosets than there are cosets. This results in small libraries for synthesis. However, there is a price to pay for such small library. Indeed, if the chain of subgroups considered has length n, then the length of the synthesized cascade is $2^n - 1$ (instead of n as in the single coset synthesis).

The subgroup $\mathbf{S}_{a-1}$ partitions its supergroup $\mathbf{S}_a$ into only two double cosets, a small one of size $(a-1)!$ and a large one of size $(a-1)!(a-1)$. Therefore, a double coset approach using the Maslov–Dueck subgroup chain (1) needs only 2^w library elements. However, a synthesized cascade can be $2^{2^w} - 1$ gates long. Therefore this subgroup chain is not a good choice in combination with double coset synthesis.

For the problem of synthesizing all members of $\mathbf{S}_8$, Van Rentergem, De Vos and Storme [23] have chosen the double cosets of the above mentioned subgroup obeying $P = A$. They conclude that, for synthesizing all 40,320 members of $\mathbf{S}_8$, they need a library of only $(4+2+1)+1 = 8$ elements. These suffice to synthesize an arbitrary member of $\mathbf{S}_8$ by a cascade with length of seven or less. For an arbitrary circuit width w, synthesis of all $(2^w)!$ members of $\mathbf{S}_{2^w}$ needs a library of $2^w - 1$ elements.

Figure 1b illustrates one step of the procedure: the 24 elements of $\mathbf{S}_4$ are fabricated by means of the 4 elements of its subgroup $\mathbf{S}_2{\times}\mathbf{S}_2$ plus the representatives of the two other double cosets in which $\mathbf{S}_4$ is partitioned by $\mathbf{S}_2{\times}\mathbf{S}_2$. Figure 2a shows how an arbitrary member g of $\mathbf{S}_{16}$ is decomposed with the help of two members (b_1 and b_2) of $\mathbf{S}_8{\times}\ \mathbf{S}_8$ and one representative a of the double coset of g. Van Rentergem et al. have demonstrated that it is always possible to construct a representative that is a control gate, i.e. a gate satisfying

$$\begin{aligned} P &= f(B, C, ...) \oplus A \\ Q &= B \\ R &= C \\ ... &\quad ... \ , \end{aligned}$$

where f (called the control function) is an arbitrary Boolean function. The control gates [24] of width w form a subgroup isomorphic to $\mathbf{S}_2^{2^{w-1}}$ of order $2^{2^{w-1}}$.

We illustrate the Van Rentergem procedure with the example of Table 4. It is the truth table of the `MILLER` gate, which is considered as a benchmark [17]. It obeys the following (non-linear) law:

$$\begin{aligned} P &= AB \oplus AC \oplus BC \\ Q &= A \oplus B \oplus AB \oplus AC \oplus BC \\ R &= A \oplus C \oplus AB \oplus AC \oplus BC \ . \end{aligned}$$

Figure 3a gives the result of repeated application of the procedure, until all subcircuits are member of $\mathbf{S}_2$, i.e. are equal to either the 1-bit identity gate or the 1-bit inverter. The nested schematic can easily be translated into a linear chain of control gates, i.e. the conventional way of writing down a reversible circuit: Figure 3b. This circuit consists of four `NOT` gates and seven `CONTROLLED CONTROLLED NOT` gates. We now introduce a cost function, called 'gate cost': we assign to each control gate a unitary cost (whatever the number of variables in the control function). The gate cost of Circuit 3b thus is 11. Note that the double coset approach ends up with a chain of cost $\frac{2}{3}\,4^w - \frac{5}{3}$ or less. Figure 3b can be further simplified, yielding Figure 3c of only 5 cost units.

As a second example, we synthesize the linear circuit of Table 1b. The result is shown in Figure 4a. We see, that although the overall circuit is linear, its synthesis contains many non-linear parts. If we like to synthesize it with exclusively linear parts, we have to choose a subgroup of

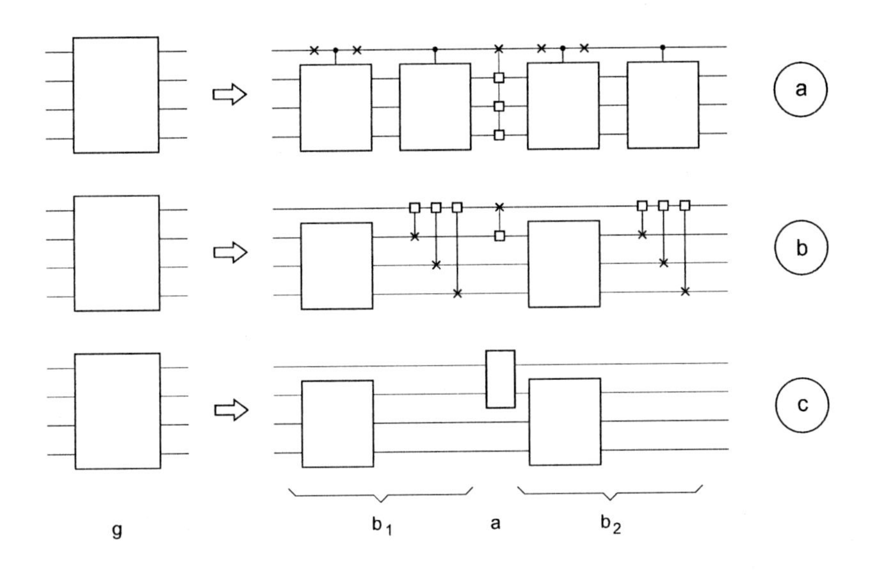

Fig. 2. An arbitrary circuit g decomposed as b_1ab_2 with the help of double cosets generated by (a) the group $\mathbf{S}_{16}$ and its subgroup $\mathbf{S}_8\times\ \mathbf{S}_8$, (b) the group $AGL(4,2)$ and its subgroup $AGL(3,2)$:$\mathbf{S}_2^3$, and (c) the group $\mathbf{S}_4$ and its subgroup $\mathbf{S}_3$.

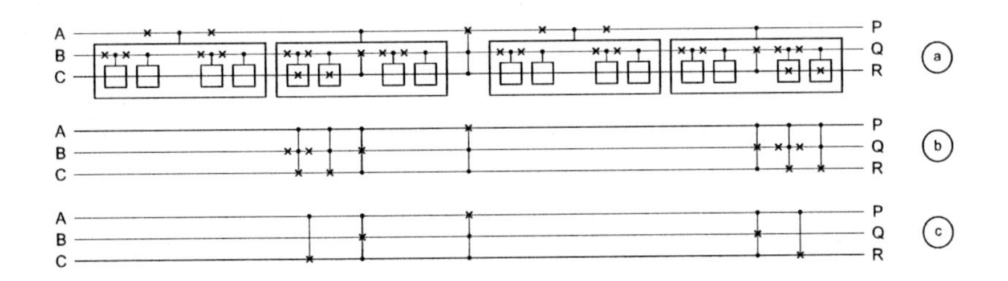

Fig. 3. Synthesis of the `MILLER` gate: (a) written as nested conditional gates, (b) written as a linear chain of conditional gates, and (c) after simplification.

Table 4. Truth table of `MILLER` gate.

ABC	PQR
0 0 0	0 0 0
0 0 1	0 0 1
0 1 0	0 1 0
0 1 1	1 0 0
1 0 0	0 1 1
1 0 1	1 0 1
1 1 0	1 1 0
1 1 1	1 1 1

$AGL(w,2)$. Just like before, we choose the subgroup of all circuits which satisfy $P = A$. This subgroup is isomorphic to the semi-direct product group $AGL(w-1,2)$ $:\mathbf{S}_2^{w-1}$ of order $2^{(w^2+w-2)/2}(w-1)!_2$. Whereas the subgroup $\mathbf{S}_{2^{w-1}} \times \mathbf{S}_{2^{w-1}}$ partitions its supergroup $\mathbf{S}_{2^w}$ into $2^{w-1}+1$ double cosets, the subgroup $AGL(w-1,2)$ $:\mathbf{S}_2^{w-1}$ partitions its supergroup $AGL(w,2)$ into only 3 double cosets, whatever the value of w. Two double cosets are small (one containing all linear reversible circuits obeying $P = A$, the other all those obeying $P = 1 \oplus A$) and one is large (containing all linear circuits obeying $\frac{\partial P}{\partial B} + \frac{\partial P}{\partial C} + ... \neq 0$). Figure 2b shows the decomposition of an arbitrary linear circuit g of $AGL(4,2)$ with the help of two members (b_1 and b_2) of $AGL(3,2)$:$\mathbf{S}_2^3$ and one representative a of the double coset of g. All control functions $f(A)$ inside b_1 and b_2 equal either 0 or A. The representative a can always be chosen from the simple set of Figure 5. These three representatives are of the form of the control gate in Figure 6, where the control function $f(B)$ equals either 0, or B, or 1. We note there exist a total of four Boolean functions of a single Boolean variable B (i.e. 0, B, $1 \oplus B$, and 1). In Figure 5 only those functions appear which are compact Maitra functions [23] [25].

Applying the 'linear' procedure again and again (starting from an arbitrary linear circuit of width 3) leads to the following chain of subgroups:

$$AGL(3,2) \supset AGL(2,2):\mathbf{S}_2^2 \supset AGL(1,2):\mathbf{S}_2^3 = \mathbf{S}_2^4 \supset \mathbf{I}\ ,$$

with subsequent orders

$$1,344 > 96 > 16 > 1\ .$$

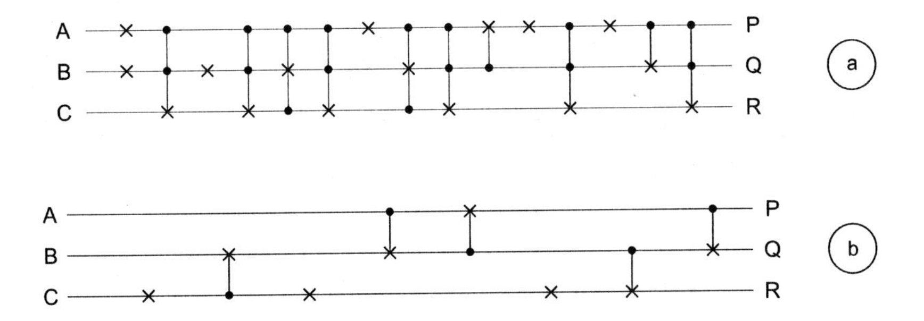

Fig. 4. Synthesis of a linear circuit: (a) written as a chain of linear as well non-linear gates, (b) written as a chain of exclusively linear gates.

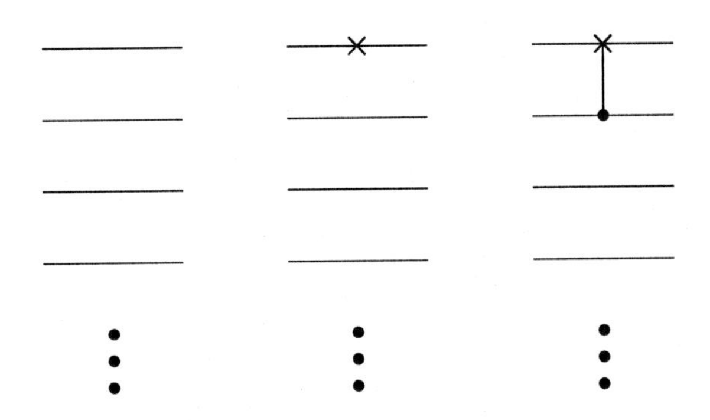

Fig. 5. Representatives of the three double cosets in which $AGL(w, 2)$ is partitioned by means of its subgroup $AGL(w-1, 2)$:$\mathbf{S}_2^{w-1}$.

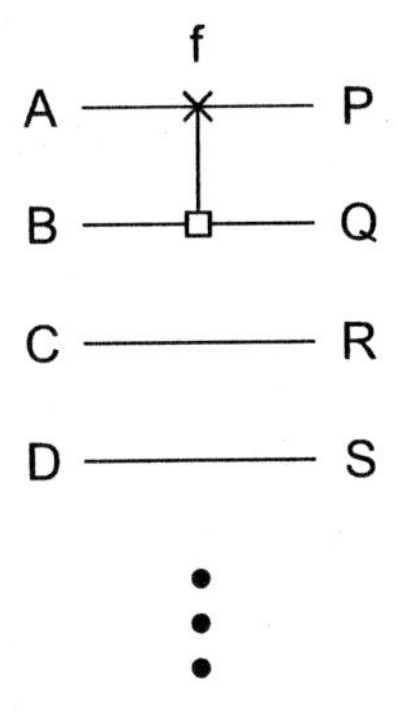

Fig. 6. Simple control gate with a single controlling bit.

Applying the new procedure to Table 1b leads to the synthesis in Figure 4b. Note that we thus end up with a chain with gate cost equal to $3 \times 2^w - 2w - 3$ or less.

We finally note there exists also a specialized synthesis method for exchangers: see Figure 2c. As explained in Section 2, the exchangers form a group isomorphic to $\mathbf{S}_w$. Those exchangers, that fulfill $P = A$ form a subgroup isomorphic to $\mathbf{S}_{w-1}$. This subgroup partitions the supergroup in only two double cosets. The representative of the small double coset (size $(w-1)!$) is the w-bit identity gate. The representative of the large double coset (size $(w-1)!(w-1)$) is the exchanger exchanging the bits A and B. Our synthesis method allows to decompose an arbitrary permutations of w wires into $2^{w-1} - 1$ (or less) neighbour exchangers.

5 More double cosets

In Section 3 we have investigated the so-called left coset space $\mathbf{H} \setminus \mathbf{G}$ and right coset space $\mathbf{G} \;/\; \mathbf{H}$. The former is the set of left cosets in which $\mathbf{G}$ is partitioned by its subgroup $\mathbf{H}$, whereas the latter is the set of right cosets in which $\mathbf{G}$ is partitioned by $\mathbf{H}$. In Section 4 we investigated the so-called double coset space $\mathbf{H} \setminus \mathbf{G} \;/\; \mathbf{H}$. This is the set of double cosets in which $\mathbf{G}$ is partitioned by its subgroup $\mathbf{H}$. We now go one step further: we take two different subgroups (say $\mathbf{H}$ and $\mathbf{K}$) of the supergroup $\mathbf{G}$ and define the double coset of an arbitrary element a of $\mathbf{G}$ as the set of all elements of $\mathbf{G}$ which can be written as $b_1 a b_2$, where b_1 is member of the subgroup $\mathbf{H}$ and b_2 is member of the subgroup $\mathbf{K}$. Again, the number of double cosets is difficult to predict. It is some number between $\frac{g}{hk}$ and $\frac{g}{l}$, where h is the order of $\mathbf{H}$, where k is the order of $\mathbf{K}$, and where l is the order of the subgroup formed by the intersection of $\mathbf{H}$ and $\mathbf{K}$. Note that $1 \leq l \leq \gcd(h,k)$.

In a first approach, De Vos et al. [26] choose for $\mathbf{K}$ a subgroup conjugate to the subgroup $\mathbf{H}$. This means that $\mathbf{K}$ equals $c\mathbf{H}c^{-1}$, with c a member of the supergroup $\mathbf{G}$. As a result we have equal orders: $k = h$. The double coset space $\mathbf{H} \setminus \mathbf{G} \;/\; \mathbf{K}$ has the same number of double cosets as the double coset space $\mathbf{H} \setminus \mathbf{G} \;/\; \mathbf{H}$. Its application nevertheless has an advantage: whatever the subgroup $\mathbf{H}$, the double coset $\mathbf{H}i\mathbf{H}$ of the identity element i is (one of) the smallest double coset(s) in the double coset space $\mathbf{H} \setminus \mathbf{G} \;/\; \mathbf{H}$. Therefore, the probability that the cheap gate i is the representative of the double coset $\mathbf{H}a\mathbf{H}$ of a given circuit a is small. On the contrary, by clever choice of the gate c, we can assure that the double coset $\mathbf{H}i\mathbf{K}$ is (one of) the largest double coset(s) of $\mathbf{H} \setminus \mathbf{G} \;/\; \mathbf{K}$, resulting in a high probability for cheap synthesis of a.

In a second approach, we choose $\mathbf{K}$ completely different from $\mathbf{H}$. The hunt for the optimal choice of the couple ($\mathbf{H}$, $\mathbf{K}$) is still open. De Vos and Van Rentergem [27] give a promising candidate in the case $\mathbf{G} = \mathbf{S}_{2^w}$, i.e. $\mathbf{H}$ isomorphic to $\mathbf{S}^2_{2^{w-1}}$ and $\mathbf{K}$ isomorphic to $\mathbf{S}_{2^{w-1}}$. E.g. in the case $\mathbf{G} = \mathbf{S}_8$, this gives $\mathbf{H}$ isomorphic to $\mathbf{S}_4^2$ and $\mathbf{K}$ isomorphic to $\mathbf{S}_4$, thus $g = 40,320$, $h = 576$, and $k = 24$.

6 Experimental prototypes

We can use either left cosets, or right cosets, or double cosets in the synthesis procedure; we can choose one subgroup or another. Whatever choices we make, we obtain a procedure for synthesizing an arbitrary gate by cascading a small number of standard cells from a limited library. By appropriate choice of the representatives of the (double) cosets, we can see to it that all building blocks in the library are member of either of the two following special subgroups:

- the subgroup of exchangers and
- the subgroup of control gates.

The former group has order $w!$ and is already discussed in Section 3. The latter group is discussed in Section 4.
Among the control gates, we note three special elements:

- If f is identically zero, then P is always equal to A. Then the gate is the identity gate i.
- If f is identically one, then P always equals $1 \oplus A$. Then the gate is an inverter or `NOT` gate: $P = \overline{A}$.
- If $f(B, C, D, ...)$ equals the $(w-1)$-bit `AND` function $BCD...$, then the gate is called the `CONTROLLED`$^{w-1}$ `NOT` gate or `TOFFOLI` gate. The former name is explained as follows: whenever $BCD...$ equals 0, then P simply equals A; but whenever $BCD...$ equals 1, then P equals `NOT` A.

For physical implementation, dual logic is very convenient. It means that any logic variable X is represented by two physical quantities, the first representing X itself, the other representing `NOT` X. Thus, e.g. the physical gate realizing logic gate of Table 1a has four physical inputs: A, `NOT` A, B, and `NOT` B, or, in short-hand notation: A, $\overline{A}$, B, and $\overline{B}$. It also has four physical outputs: P, $\overline{P}$, Q, and $\overline{Q}$. Such approach is common in electronics, where it is called dual-line or dual-rail electronics. Also some quantum computers make use of dual-rail qubits [28]. As a result, half of the input pins are at logic 0 and the other half at logic 1, and analogous for the output pins. In this way, dual electronics is physically conservative: the number of 1s at the output equals the number of 1s at the input (i.e. equals w), even if the truth table of the logic gate is not conservative. As a result, we get the advantages of conservative logic, without having to restrict ourselves to conservative logic.

Dual-line hardware allows very simple implementation of the inverter. It suffices to interchange its two physical lines in order to invert a variable, i.e. in order to hardwire the `NOT` gate. Conditional `NOT`s are `NOT` gates which are controlled by switches. A first example is the `CONTROLLED NOT` gate:

$$P = A$$
$$Q = A \oplus B \ .$$

These logic relationships are implemented into the physical world as follows:

- output P is simply connected to input A,
- output $\overline{P}$ is simply connected to input $\overline{A}$,
- output Q is connected to input B if $A = 0$, but connected to $\overline{B}$ if $A = 1$, and
- output $\overline{Q}$ is connected to input $\overline{B}$ if $A = 0$, but connected to B if $A = 1$.

The last two implementations are shown in Figure 7a. In the figure, the arrow heads show the position of the switches if the accompanying label is 1. A second example is the `CONTROLLED CONTROLLED NOT` gate or `TOFFOLI` gate:

$$P = A$$
$$Q = B$$
$$R = AB \oplus C \ .$$

Its logic relationships are implemented into physical world as follows:

- output P is simply connected to input A,
- output $\overline{P}$ is simply connected to input $\overline{A}$,
- output Q is simply connected to input B,
- output $\overline{Q}$ is simply connected to input $\overline{B}$,
- output R is connected to input C if either $A = 0$ or $B = 0$, but connected to $\overline{C}$ if both $A = 1$ and $B = 1$, and
- output $\overline{R}$ is connected to input $\overline{C}$ if either $A = 0$ or $B = 0$, but connected to C if both $A = 1$ and $B = 1$.

The last two implementations are shown in Figure 7b. Note that in both Figures 7a and 7b, switches always appear in pairs, of which one is closed whenever the other is open and vice versa. It is clear that the above design philosophy can be extrapolated to a control gate with arbitrary control function f. Suffice it to wire a square circuit like in Figures 7a and 7b, with the appropriate series and parallel connection of switches. Note that, whenever the change of a controlling voltage opens a current-conducting switch, it simultaneously provides an alternate path for the current. This is a necessary condition for reversibility.

Now that we have an implementation approach, we can realize any reversible circuit in hardware. We will demonstrate here some examples of implementation into electronic chip. In electronic circuits, a switch is realized by the use of a so-called transmission gate, i.e. two MOS-transistors in parallel (one n-MOS transistor and one p-MOS transistor). As an example [6], Figure 8 shows a 4-bit ripple adder, implemented in 2.4 μm standard c-MOS technology, consisting of eight `CONTROLLED NOT`s and eight `CONTROLLED CONTROLLED NOT`s, and thus of a total of 192 transistors. This prototype chip was fabricated in 1998. A second example [7] (Figure 9) was fabricated in 2000, in submicron technology: a 4-bit carry-look-ahead adder, implemented in 0.8 μm standard c-MOS technology, containing four `CONTROLLED NOT`s, four control gates of width $w = 3$, and one complex control gate of width $w = 13$. It contains a total of 320 transistors.

Switches not only can decide whether an input variable is inverted or not. We can apply switches also in order to decide whether two input variables are swapped or not. This concept leads to the `CONTROLLED SWAP` gate or `FREDKIN` gate [11]:

$$\begin{aligned} P &= A \\ Q &= B \oplus AB \oplus AC \\ R &= C \oplus AB \oplus AC \ . \end{aligned}$$

Figure 7c shows the physical implementation. The reader will easily extrapolate the design philosophy to reversible logic gates of width $w = w_1 + w_2$, where w_1 controlling bits decide, by means of a control function f, whether the w_2 controlled bits are subjected to a given selective reversible gate or not. As there exist $2^{2^{w_1}}$ possible control functions and $2^{w_2} w_2!$ possible selective gates, we can construct a library of $2^{2^{w_1}} \times 2^{w_2} w_2!$ such building blocks.

An application [8] (Figure 10) is a 1-bit (full) adder, implemented in 0.35 μm standard c-MOS technology, containing three `CONTROLLED NOT`s and one `FREDKIN` gate. It contains a total of 40 transistors. The prototype chip was fabricated in 2004. The reader will observe that full-

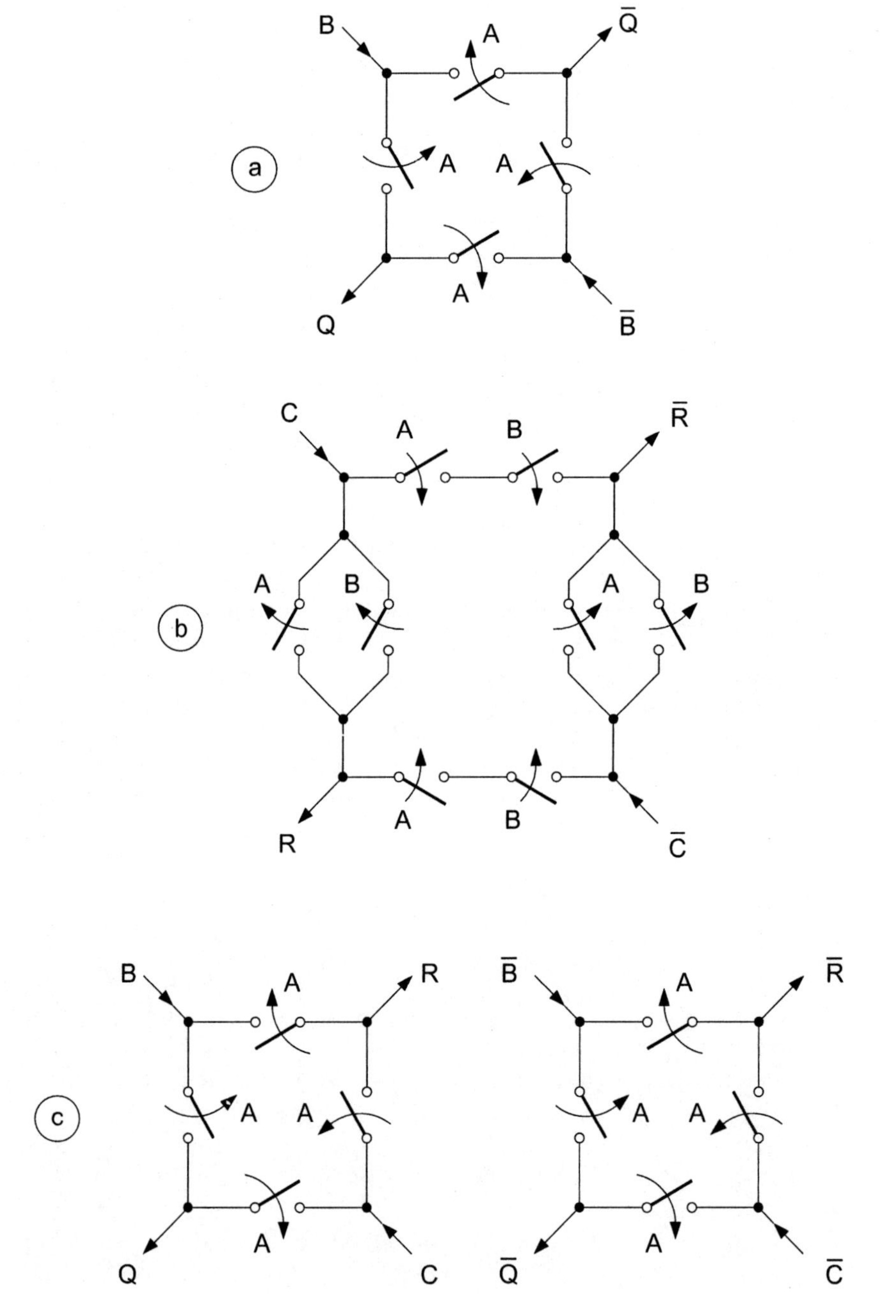

Fig. 7. Schematic for (a) `CONTROLLED NOT` gate, (b) `CONTROLLED CONTROLLED NOT` gate, and (c) `CONTROLLED SWAP` gate [8].

Fig. 8. Microscope photograph (140 μm $\times$ 120 μm) of 2.4-μm 4-bit reversible ripple adder [6].

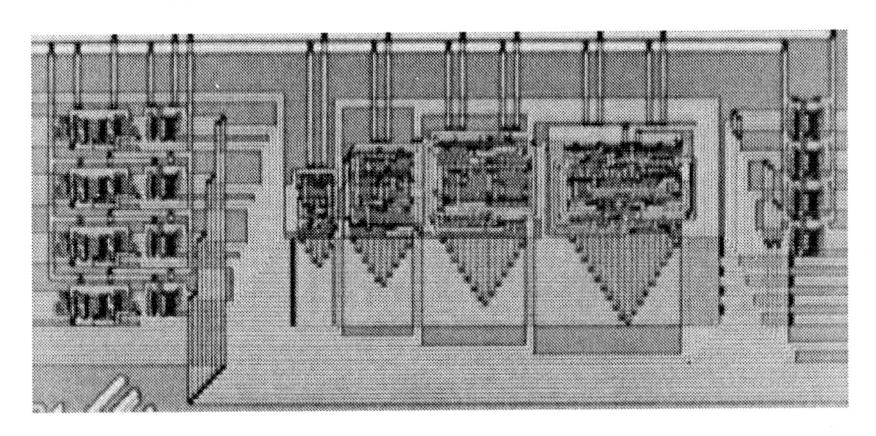

Fig. 9. Microscope photograph (610 μm $\times$ 290 μm) of 0.8-μm 4-bit reversible carry-look-ahead adder [7].

custom prototyping at university follows Moore's law, with a couple of years delay with respect to industry. Indeed, many commercial chips nowadays use 0.18 or 0.13 μm transistors. Some companies even have entered the nanoscale era, by introducing 90 nm products (and smaller) to the market.

The continuing shrinking of the transistor sizes leads to a continuing decrease of the energy dissipation per computational step. This heat generation Q is of the order of magnitude of CV_t^2, where V_t is the threshold voltage of the transistors and C is the total capacitance of the capacitors in the logic gate [29]. We have C of the order of magnitude of $\epsilon_0\epsilon\,\frac{WL}{t}$, where W, L, and t are the width, the length, and the oxide thickness of the transistors, whereas ϵ_0 is the permittivity of vacuum ($8.85\ \times 10^{-12}$ F/m) and ϵ is the dielectric constant of the gate oxide. Table 5 gives some typical numbers. The dielectric constant is taken 3.9 (silicondioxide SO_2). We see how Q becomes smaller and smaller. However, dissipation in electronic circuits still is about four orders of magnitude in excess of the Landauer quantum $kT\log(2)$, which amounts (for $T = 300$ K) to about 3×10^{-21} J or 3 zeptojoule.

Further shrinking of W and L and further reduction of V_t ultimately will lead to a Q value in the neighbourhood of $kT\log(2)$. That day, digital electronics will have good reason to be reversible. This, however, does not mean that the reversible MOS circuits are useless today. Indeed, as

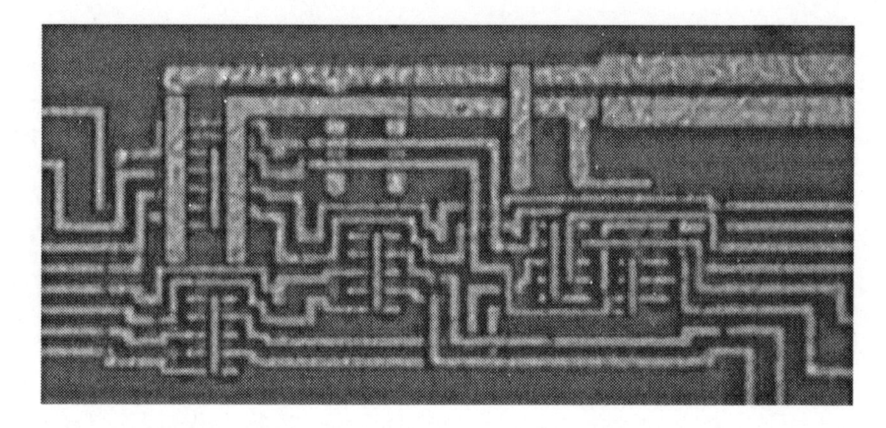

Fig. 10. Microscope photograph (80 μm $\times$ 37 μm) of 0.35-μm 1-bit reversible full adder.

Table 5. Moore's law for dimensions W, L, and t, and for threshold voltage V_t, as well as for resulting capacitance C and heat dissipation Q.

technology (μm)	W (μm)	L (μm)	t (nm)	V_t (V)	C (fF)	Q (fJ)
2.4	2.4	2.4	42.5	0.9	46.8	38
0.8	0.8	2.0	15.5	0.75	3.6	2.0
0.35	0.35	0.5	7.4	0.6	0.82	0.30

they are a reversible form of pass-transistor topology, they are particularly suited for adiabatic addressing [30], leading to substantial power saving. In subsequent gates, switches are opening and closing, one after the other, like dominoes, transferring information from the inputs to the outputs [31]. Figure 11 shows an example of a quasi-adiabatic experiment. We see two transient signals: one of the input variables and one of the resulting output bits. In practice, such procedure leads to a factor of about 10 in power reduction [29]. The reduction of the power dissipation is even more impressive if standard c-MOS technology is replaced by SOI (silicon-on-insulator) technology. Indeed, in the latter process, the threshold voltage V_t can be controlled better, such that low-V_t technologies are possible.

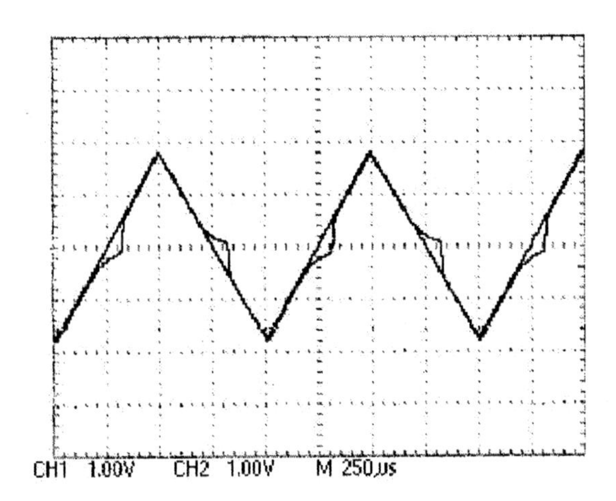

Fig. 11. Oscilloscope view of 0.35 μm full adder [8].

7 Conclusion

We have demonstrated that subgroup chains of the form $\mathbf{S}_8 \supset AGL(3,2) \supset \mathbf{S}_3 : \mathbf{S}_2^3 \supset \mathbf{S}_3 \supset \mathbf{S}_1$ divide the group of reversible logic circuits in sets of elements that are less and less suited as building-blocks for design of reversible circuits three units wide. We have shown how cosets and double cosets are particularly helpful for synthesizing arbitrary reversible circuits. In particular, we have demonstrated the power of subgroup chains of the form $\mathbf{S}_8 \supset \mathbf{S}_7 \supset \mathbf{S}_6 \supset \mathbf{S}_5 \supset \mathbf{S}_4 \supset \mathbf{S}_3 \supset \mathbf{S}_2 \supset \mathbf{S}_1$ and the form $\mathbf{S}_8 \supset \mathbf{S}_4^2 \supset \mathbf{S}_2^4 \supset \mathbf{S}_1^8$. For arbitrary linear reversible circuits, the following chain form is useful: $AGL(3,2) \supset AGL(2,2) :\mathbf{S}_2^2 \supset AGL(1,2) :\mathbf{S}_2^3$. All synthesis methods, described for circuit width w equal to 3, are immediately applicable to larger w, with the help of similar subgroup chains of $\mathbf{S}_{2^w}$ and $ALG(w,2)$.

Control gates belong to a subgroup isomorphic to $\mathbf{S}_2^{2^{w-1}}$. They form the building blocks for hardware implementation. Silicon prototypes of adder circuits, in three different standard technologies (2.4 μm, 0.8 μm, and 0.35 μm), have been fabricated and tested.

References

1. I. Markov, An introduction to reversible circuits, *Proceedings of the International Workshop on Logic and Synthesis*, Laguna Beach (May 2003), pp. 318 - 319.
2. M. Frank, Introduction to reversible computing: motivation, progress, and challenges, *Proceedings of the 2005 Computing Frontiers Conference*, Ischia (May 2005), pp. 385 - 390.
3. A. De Vos, Lossless computing, *Proceedings of the I.E.E.E. Workshop on Signal Processing*, Poznań (October 2003), pp. 7 - 14.
4. B. Hayes, Reverse engineering, *American Scientist* **94** (March–April 2006), pp. 107 - 111.
5. R. Feynman, Quantum mechanical computers, *Optics News* **11** (1985), pp. 11 - 20.
6. B. Desoete, A. De Vos, M. Sibiński, and T. Widerski, Feynman's reversible logic gates, implemented in silicon, *Proceedings of the 6 th International Conference on Mixed Design of Integrated Circuits and Systems*, Kraków (June 1999), pp. 497 - 502.
7. B. Desoete and A. De Vos, A reversible carry-look-ahead adder using control gates, *Integration, the V.L.S.I. Journal* **33** (2002), pp. 89 - 104.
8. Y. Van Rentergem and A. De Vos, Optimal design of a reversible full adder, *International Journal of Unconventional Computing* **1** (2005), pp. 339 - 355.
9. W. Scott, Group theory, Dover Publications (New York, 1964).
10. P. Hall, The theory of groups, AMS Chelsea Publishing (Providence, 1968).
11. E. Fredkin and T. Toffoli, Conservative logic, *International Journal of Theoretical Physics* **21** (1982), pp. 219 - 253.
12. K. Patel, I. Markov, and J. Hayes, Optimal synthesis of linear reversible circuits, *Proceedings of the 13 th International Workshop on Logic and Synthesis*, Temecula (June 2004), pp. 470 - 477.
13. S. Aaronson and D. Gottesman, Improved simulation of stabilizer circuits, *Physical Review A* **70** (2004), 052328.
14. L. Wang and A. Almaini, Optimisation of Reed–Muller PLA implementations, *I.E.E. Proceedings – Circuits, Devices and Systems* **149** (2002), pp. 119 - 128.

15. P. Kerntopf, On universality of binary reversible logic gates, *Proceedings of the 5 th Workshop on Boolean Problems*, Freiberg (September 2002), pp. 47 - 52.
16. A. De Vos and L. Storme, r-Universal reversible logic gates, *Journal of Physics A: Mathematical and General* **37** (2004), pp. 5815 - 5824.
17. D. Maslov and G. Dueck, Reversible cascades with minimal garbage, *I.E.E.E. Transactions on Computer-Aided Design of Integrated Circuits and Systems* **23** (2004), pp. 1497 - 1509.
18. Y. Van Rentergem, A. De Vos, and K. De Keyser, Using group theory in reversible computing, *I.E.E.E. World Congress on Computational Intelligence*, Vancouver, 16-21 July 2006.
19. Y. Van Rentergem, A. De Vos, and K. De Keyser, Six synthesis methods for reversible logic, *Open Systems & Information Dynamics*, accepted for publication.
20. A. Kerber, Representations of permutation groups I, *Lecture Notes in Mathematics*, volume 240, Springer Verlag (Berlin, 1970), pp. 17-23.
21. G. James and A. Kerber, The representation theory of the symmetric group, *Encyclopedia of Mathematics and its Applications*, volume 16 (1981), pp.15-33.
22. A. Jones, A combinatorial approach to the double cosets of the symmetric group with respect to Young subgroups , *European Journal of Combinatorics* **17** (1996), pp. 647-655.
23. Y. Van Rentergem, A. De Vos, and L. Storme, Implementing an arbitrary reversible logic gate , *Journal of Physics A: Mathematical and General* **38** (2005), pp. 3555 - 3577.
24. A. De Vos, B. Raa, and L. Storme, Generating the group of reversible logic gates , *Journal of Physics A: Mathematical and General* **35** (2002), pp. 7063 - 7078.
25. R. Minnick, Cutpoint cellular logic , *I.E.E.E. Transactions on Electronic Computers* **13** (1964), pp. 685 - 698.
26. A. De Vos, Y. Van Rentergem, and K. De Keyser, The decomposition of an arbitrary reversible logic circuit , *Journal of Physics A : Mathematical and General* **39** (2006), pp. 5015 - 5035.
27. A. De Vos and Y. Van Rentergem, Double coset spaces for reversible computing , *Proceedings of the 7 th International Workshop on Boolean Problems*, Freiberg (September 2006).
28. I. Chuang and Y. Yamamoto, The dual-rail quantum bit and quantum error correction , *Proceedings of the 4 th Workshop on Physics and Computation*, Boston (November 1996), pp. 82 - 91.
29. A. De Vos and Y. Van Rentergem, Energy dissipation in reversible logic addressed by a ramp voltage , *Proceedings of the 15 th International PATMOS Workshop*, Leuven (September 2005), pp. 207 - 216.
30. P. Patra and D. Fussell, On efficient adiabatic design of MOS circuits , *Proceedings of the 4 th Workshop on Physics and Computation*, Boston (November 1996), pp. 260 - 269.
31. M. Alioto and G. Palumbo, Analysis and comparison on full adder block in submicron technology , *I.E.E.E. Transactions on Very Large Scale Integration Systems* **10** (2002), pp. 806 - 823.

32. P. Cameron, R. Solomon, and A. Turull, Chains of subgroups in symmetric groups , *Journal of Algebra* **127** (1989), pp. 340 - 352.
33. P. Cameron, Permutation groups , *London Mathematical Society Student Texts* **45** (1999), Cambridge University Press, Cambridge.

Appendix

If $\mathbf{G}_1$ (with order g_1) is a subgroup of the group $\mathbf{G}$ (with order g), then $\mathbf{G}_1$ partitions $\mathbf{G}$ into $\frac{g}{g_1}$ classes, all of the same size g_1.

Cosets are very handy in synthesis. Assume we want to make an arbitrary element of the group $\mathbf{G}$ in hardware. Instead of solving this problem for each of the g cases, we only synthesize the g_1 gates b of $\mathbf{G}_1$ and a single representative r_i of each other left coset ($1 \leq i \leq \frac{g}{g_1} - 1$). If we can make each of these $\frac{g}{g_1} - 1 + g_1$ gates, we can make all the others by merely making a short cascade br_i. If we cleverly choose the subgroup $\mathbf{G}_1$, we can guarantee that $\frac{g}{g_1} - 1 + g_1$ is much smaller than g. We call the set of $\frac{g}{g_1} - 1 + g_1$ building-blocks the library for synthesizing the g gates of $\mathbf{G}$. What is a clever choice of $\mathbf{G}_1$? The library will be as small as possible if $d(\frac{g}{g_1} - 1 + g_1)/dg_1$ is zero, i.e. if $g_1 = \sqrt{g}$, leading to the minimum size $2\sqrt{g} - 1$. As an example, we consider the case where $\mathbf{G}$ equals the symmetric group $\mathbf{S}_8$. As $g = 40{,}320$, the optimum subgroup should have order $\sqrt{40,320} \approx 200.7$. Of course, a subgroup order g_1 has to be an integer. Moreover, because of Lagrange's Theorem, the only g_1-values which can exist have to be divisors of g. Thus we have to choose a subgroup $\mathbf{G}_1$ with order 'close to' the square root of the order g of $\mathbf{G}$, e.g. $g_1 = 192$. Figure 12 shows another example, i.e. $\mathbf{S}_4$ with $g = 4! = 24$. We see the curve $g_1 + \frac{24}{g_1} - 1$ with minimum value $2\sqrt{g} - 1$ equal to about 8.8, at $g_1 = \sqrt{g} \approx 4.9$. The crosses indicate really existing subgroups. Several of them are located close to the curve's minimum. They have order g_1 either equal to 4 or to 6.

One can go one step further: one can synthesize the g_1 elements of $\mathbf{G}_1$ by choosing a subgroup $\mathbf{G}_2$ of $\mathbf{G}_1$. If g_2 is the order of $\mathbf{G}_2$, then $\mathbf{G}_2$ will partition $\mathbf{G}_1$ into g_1/g_2 cosets. We therefore will need $\frac{g_1}{g_2} - 1 + g_2$ building-blocks to synthesize all g_1 elements of $\mathbf{G}_1$. Thus, for synthesizing all g elements of $\mathbf{G}$ as a cascade of three building-blocks, a library of $\frac{g}{g_1} - 1 + (\frac{g_1}{g_2} - 1 + g_2) = (\frac{g}{g_1} - 1) + (\frac{g_1}{g_2} - 1) + g_2$ is needed. This library will have minimum size if both $\partial(\frac{g_1}{g_2} + \frac{g}{g_1} + g_2 - 2)/\partial g_1$ and $\partial(\frac{g_1}{g_2} + \frac{g}{g_1} + g_2 - 2)/\partial g_2$ are equal to zero. This yields $g_1 = g^{2/3}$ and $g_2 = g^{1/3}$. We thus need a subgroup chain

$$\mathbf{G} \supset \mathbf{G}_1 \supset \mathbf{G}_2 \supset \mathbf{I} \ ,$$

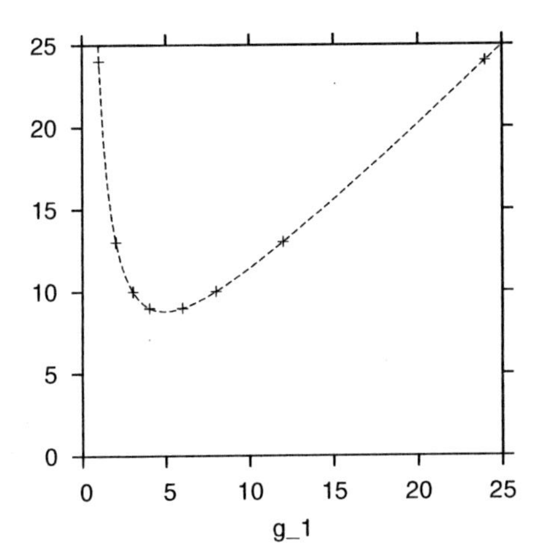

Fig. 12. Library size generated by the left/right cosets of an order-g_1 subgroup of $\mathbf{S}_4$.

with subsequent orders

$$g > g^{2/3} > g^{1/3} > 1 \ .$$

Proceeding in the same way, one can construct a subgroup chain

$$\mathbf{G} \supset \mathbf{G}_1 \supset \mathbf{G}_2 \supset ... \supset \mathbf{G}_{n-1} \supset \mathbf{G}_n = \mathbf{I} \ .$$

The number n is called the length of the chain. The resulting library has size $(\frac{g}{g_1} - 1) + (\frac{g_1}{g_2} - 1) + ... + (\frac{g_{n-1}}{g_n} - 1) + g_n = \frac{g}{g_1} + \frac{g_1}{g_2} + ... + \frac{g_{n-1}}{1} - n + 1$. The optimal sequence of subgroup orders

$$g > g_1 > g_2 > ... > g_{n-1} > g_n = 1 \ ,$$

is

$$g > g^{(n-1)/n} > g^{(n-2)/n} > ... > g^{1/n} > 1 \ ,$$

giving rise to a library size of $ng^{1/n} - n + 1$. Any element of $\mathbf{G}$ can then be decomposed into a cascade with length n. Thus the orders of the subsequent subgroups have to form (as well as possible) a geometric series. Of course, with increasing n, it becomes more and more improbable that, for a given group $\mathbf{G}$, such a subgroup chain actually exists. Moreover, it is impossible to increase n beyond some maximum value. Because each of the indexes g/g_1, g_1/g_2, g_2/g_3, ..., g_{n-1}/g_n has to be an integer larger than or equal to 2, we know that, for any group $\mathbf{G}$, the maximum value n_{max}, i.e. the length of the longest subgroup chain, satisfies

$$n_{max} \leq \log_2(g) \ .$$

For the special case where $\mathbf{G}$ is isomorphic to the symmetric group $\mathbf{S}_{2^w}$, thanks to a theorem by Cameron et al. [32] [33], we known that

$$n_{max} = \frac{3}{2}\, 2^w - 2 \ .$$

This, in turn, leads to a lower bound for the library size:

$$n_{max} g^{1/n_{max}} - n_{max} + 1 \ .$$

For $\mathbf{G}$ isomorphic to $\mathbf{S}_8$, we conclude that any subgroup chain has a length shorter than or equal to 10 and yields a library size larger than 19.

We remind that the Maslov–Dueck chain (1) has length $n = 2^w - 1 = 7$. Its chain of subgroup orders (2) is close to the ideal chain with $n = 7$, i.e. the geometric progression

$$\begin{aligned} 40,320 &> 40,320^{\,6/7} > 40,320^{\,5/7} > 40,320^{\,4/7} \\ &> 40,320^{\,3/7} > 40,320^{\,2/7} > 40,320^{\,1/7} > 1 \ , \end{aligned}$$

which, after approximating to the nearest integers, evaluates as

$$40,320 > 8,850 > 1,942 > 426 > 95 > 21 > 5 > 1 \ .$$

The Van Rentergem–De Vos chain (3) has length $n = w = 3$. Its chain of subgroup orders (4) is close to the ideal chain with $n = 3$, i.e. the geometric progression

$$40,320 > 40,320^{\,2/3} > 40,320^{\,1/3} > 1 \ ,$$

which, after approximating to the nearest integers, evaluates as

$$40,320 > 1,176 > 34 > 1 \ .$$

Of course, a same conslusion holds for right coset synthesis. Unfortunately, no similar theory can be constructed for double coset synthesis. Indeed, no simple formula exists for the number of double cosets in $\mathbf{H} \setminus \mathbf{G} \,/\, \mathbf{H}$, let alone in $\mathbf{H} \setminus \mathbf{G} \,/\, \mathbf{K}$ with $\mathbf{K}$ different from $\mathbf{H}$. Even in the special case, where both $\mathbf{H}$ and $\mathbf{K}$ are Young subgroups of $\mathbf{G}$, counting the double cosets is not a straightforward task.

Index

Printed in the United Kingdom
by Lightning Source UK Ltd.
112011UKS00001B/1-60